2   1.8   1   0.5   0.01 0 MA

Later *Gigantopithecus*

Hylobatidae

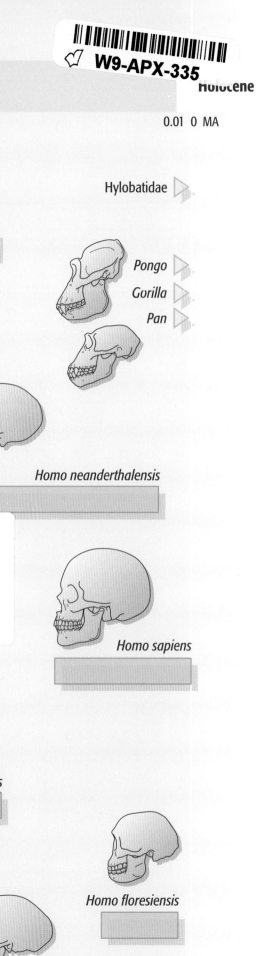

Pongo

Gorilla

Pan

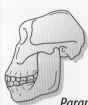

Paranthropus boisei

Homo neanderthalensis

Paranthropus robustus

Homo habilis

Homo sapiens

Homo heidelbergensis

?

Homo ergaster

Homo floresiensis

Homo erectus

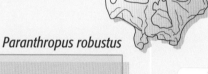

THE ORIGINS
of MAN

THE ORIGINS
of MAN

DR DOUGLAS PALMER

First published in 2007 by New Holland Publishers (UK) Ltd
London • Cape Town • Sydney • Auckland

10 9 8 7 6 5 4 3 2 1

www.newhollandpublishers.com

Garfield House, 86–88 Edgware Road, London W2 2EA, United Kingdom

80 McKenzie Street, Cape Town 8001, South Africa

14 Aquatic Drive, Frenchs Forest, NSW 2086, Australia

218 Lake Road, Northcote, Auckland, New Zealand

ISBN 978 1 84537 165 4

Editorial Director: Jo Hemmings
Commissioning Editor: Simon Pooley
Senior Editor: Charlotte Judet
Design: Peter Crump
Production: Marion Storz
Cartography: Martin Sanders and William Smuts
Illustrations: Debbie Maizels

Reproduction by Modern Age Repro House Ltd, Hong Kong
Printed and bound in Singapore by Tien Wah Press (Pte) Ltd

Cover and Preliminary pages:
Front cover: *Sahelanthropus* skull (left); Moving beyond Africa map
(middle); *Australopithecus africanus* skull (right). Spine: Laetoli footprints.
Back cover: Our cousins, the chimpanzees (left); Southen Apes of Africa map
(middle); Neanderthal reconstruction (right). Page 3: Papua New Guinea
man. Page 4–5: Reconstruction of Neanderthal hunt. Page 6: Reconstruction
of Neanderthal spear.

Author's Acknowledgements

The author wishes to thank Simon Pooley for initiating the
project, Paul Bahn (consultant) and Rod Baker (copy
editor) for lots of helpful suggestions and particularly
Charlotte Judet (editor) for seeing it through to completion
with the help of Peter Crump (designer), William Smuts
and Martin Sanders (maps). Special thanks go to Debbie
Maizels for her artwork that has made an important
contribution to the presentation of some complicated data.

Author's Dedication
To Tamsy, for 25 years of stone and bones.

Contents

Introduction

 There are some six billion humans spread around the globe today and we all belong to the same species *Homo sapiens*. The proof of our kinship is to be found in any big cosmopolitan city where peoples of different ethnic backgrounds have met and paired up to produce children. A biological species is one that can successfully breed and whose offspring can also reproduce. It does not matter which of the world's 6000 or so languages we speak or, whether one partner is an Inuit from arctic Canada and the other comes from a tropical island in the Pacific. Although originally a tropical species, our adaptability and technological know-how have allowed us to occupy a greater range of environments than any other species. The story of our species' success is a remarkable one and is the subject of this book.

If two modern humans of opposite sex feel sufficiently attracted the chances are that they can have children together, but they can only achieve this because genetically we are so similar to one another. And, since our genes evolve over time, this genetic closeness means that all living humans must share a common ancestor who lived no more than some 100,000 years ago. We now know, as Charles Darwin predicted, that our common ancestor lived in Africa and evolved from an ape-like ancestor some six or seven million years ago.

Breaking new ground, breaching old barriers

When Darwin set out on his epic voyage on the *Beagle* in 1831, such an understanding of human prehistory and origins was more or less unthinkable and likely to induce severe criticism. In the first few decades of the 19th century scientists were just beginning to break some of the cultural, religious and associated conceptual barriers, which obscured their understanding of the past. Biologists were beginning to question the idea that species were fixed and therefore 'immutable'. They could see that as more and more examples of different plants and

animals, both living and extinct (from the fossil record) were revealed and classified, the gaps between different forms were often filled by more closely related forms. And, when the prehistorical changes in life, as revealed by fossils, were arranged against their chronological appearance in the rock record,

Even in the late 1850s when Darwin was writing the *Origin of Species*, the question of the evolutionary origin of humankind was still a taboo subject. All Darwin suggested was that, in future, 'light would be thrown' on our origins, but even that comment brought enormous criticism.

Darwin was often caricatured as an ape or monkey because of his ideas about human evolution. In this late 19th century cartoon he is portrayed as a performing circus monkey bursting through the hoops of 'credulity', 'superstition' and 'ignorance', to the astonishment of a human clown. The other monkey who is holding the hoops is the French physician and philosopher, Émile Littré (1801-1881) who supported the 'irreligious' ideas of Auguste Comte, founder of positivism. Littré returned to Catholicism on his deathbed.

another pattern emerged. The deeper geologists dug into the record of successive strata exposed at the Earth's surface, the more they revealed of this pattern with its marked changes in whole groups of organisms. Overall there seemed to be a connection between primitive life forms and the oldest known strata with more advanced groups such as the mammals and birds only appearing later in the record. But how could this be interpreted?

Today, with the Darwin–Wallace theory of evolution behind us, it all seems so obvious but in the 1830s it was not. Some eminent geologists such as Charles Lyell expected that fossils of even the most advanced forms might be found in very ancient strata because all of life might have appeared at once as if by some magic wand or the hand of a deity. But Lyell and the other theologically minded scientists were forced to give way as overwhelming

The succession of rock strata and their contained fossils was being mapped out for the first time in the early decades of the 19th century by the likes of William Smith, a largely self-taught English surveyor and geologist. Superficial deposits, known as Diluvium, spread over low lying land especially in parts of southern England, were known to contain the fossil remains of large mammals such as extinct elephant and rhino species. Their presence was something of a puzzle and possibly resulted from the Noachian Flood, but by the middle decades of the 19th century it was realised that the Diluvium and its fossils resulted from recent ice ages over much of Northern Europe.

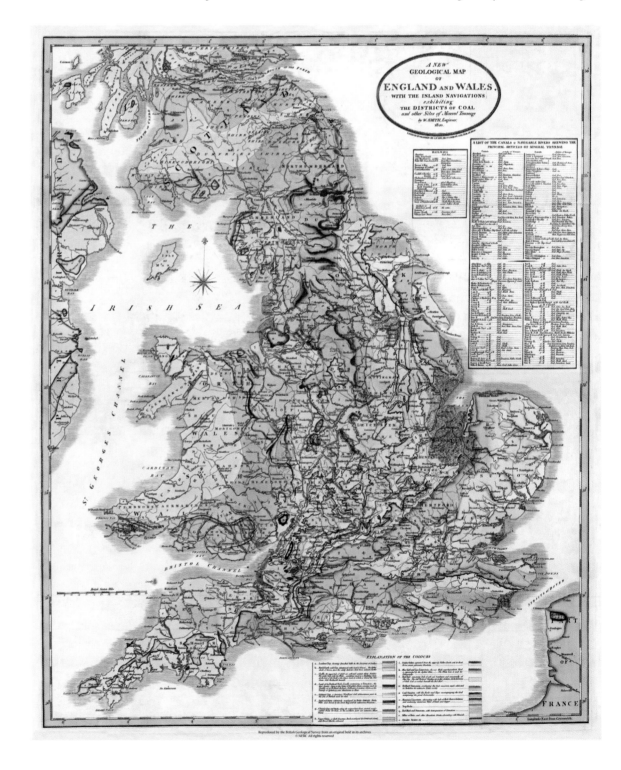

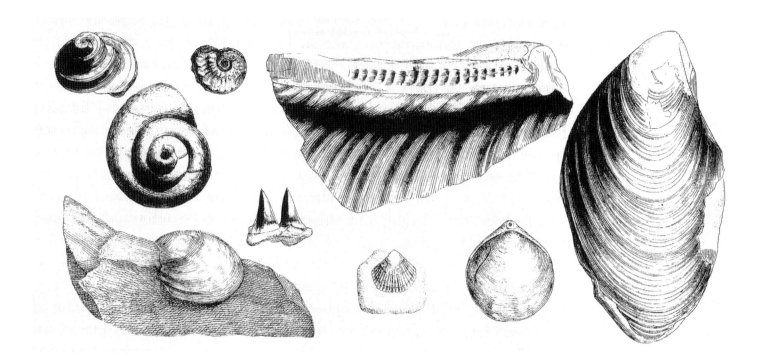

evidence appeared to show the progressive appearance of more advanced animals and plants through successive strata and therefore time.

Interpreting the evidence in the rocks

Darwin carefully avoided the issue of the rock and fossil record because he knew that the contemporary understanding was still too impoverished and problematic. He was also well aware that many scientists, such as his Cambridge mentor on geology, the Reverend Professor Adam Sedgwick, knew a lot more about rocks and fossils than he did. However, even Sedgwick and his contemporaries had a struggle with the interpretation of evidence they found in the rocks. They had expected the truth of the Creation story and that of the Flood to be revealed by the testimony of the rocks. But what they discovered was much more complex and worrying.

The total thickness of fossil-bearing strata was clearly immense – amounting to many kilometres – far too many to be explained away by any number of floods. Understanding of geological processes of rock formation and destruction told them that the formation of such a thickness of strata clearly represented an immense length of time, perhaps many millions of years. The old bible-derived chronology of the earth's formation over a mere 6,000 or so years would not do. But nobody could find an acceptable scientific way of measuring the

age of the rocks and their contained fossils. All they had was a relative chronology based on the mapping of successive strata and their characteristic fossils. The technique was pioneered in Britain by the largely self-taught English surveyor and geologist William Smith and at the same time in France by the anatomist Georges Cuvier (1769-1832) and the geologist Alexandre Brongniart (1770-1847).

Archaeopteryx – the vital find

By the 1860s and the publication of the *The Origin of Species*, most of the major divisions of geological time had been named and a reasonable sample of their characteristic fossils was generally known. Even so, Darwin shied away from getting embroiled with the fossil record and went to considerable lengths to explain why. He correctly argued that the rock record is full of gaps and so therefore is its record of past life. He knew that if his theory of evolution was to survive the test of the fossil record then strata should be found to contain fossils that demonstrated common ancestry and divergences between evolving lineages.

What are now seen as clearly separate groups such as humans and apes, or primates and insectivores, or mammals and reptiles, must have had common ancestors at some point in the past. The problem was that in the mid-19th century such fossils were not evident. Fortunately for Darwin,

William Smith was one of the first people to recognise that sequences of strata could be distinguished by their contained fossils which change over time. Here are Smith's illustrations of characteristic fossils of the Cretaceous Chalk. At this time no human remains had been found in any strata and consequently their absence seemed to support the idea of a special creation for humanity.

evolution, rather than some divine intervention, were still missing. As we shall see, even when such fossils were first found in the early decades of the 19th century they were not properly accepted for what they were. This discovery and delayed acceptance of our prehistory as a species is an important part of the story we are telling here.

The evidence gains acceptance

The discovery that it is possible to accurately measure the chronology of the geological past was even later in arriving on the scientific scene. Not until the discovery of radioactive isotopes at the beginning of the 20th century did radiometric dating become a possibility. Even then, understanding of the processes and techniques required for such measurements took decades to be refined sufficiently to become widely applicable to the dating of our human prehistory.

As we shall see, it was not until the latter part of the 20th century that the fossil and archaeological evidence for human evolution finally began to be accepted. But that evidence did not seem at first to support Darwin's claim that the origin of humankind was to be found in Africa, the home of

The 1861 discovery of a fossilised bird from Jurassic age strata in Germany provided critical support for the Darwin–Wallace theory of evolution since it still preserves some ancestral reptilian features such as a long bony tail and toothed jaws.

The Curies' discovery of radioactivity in the late 19th century eventually led to our ability to radiometrically date minerals and rocks.

however, one of the first and most spectacular fossils to demonstrate links between two major groups, the birds and the reptiles, *Archaeopteryx* – did show up in 1861. And although one of his main critics, the anatomist Richard Owen, tried to use it against Darwin, he failed. Thomas Henry Huxley, a champion of evolution, took advantage of Owen's failure and turned the tables on him to argue one of the most convincing fossil examples of evolutionary links between major groups of animals – in this instance, the reptiles and birds.

But of course the most critical fossils required for the acceptance of the idea that humans might also have been subjected to the same processes of those higher apes that are our closest relatives – the gorillas and chimpanzees. Initially, fossil evidence that was emerging seemed to support the claim of an ardent evolutionist fan of Darwin's, the German biologist Ernst Haeckel, that Asia rather than Africa was our 'Eden'. Not until the 1920s did the tide begin to turn in the favour of Africa and Darwin.

Africa confirmed, the cradle of humankind

Now, however, the evidence for an African 'Eden' is overwhelming. The fossil evidence shows that most of our extinct relatives lived in Africa and radiometric dating of the strata within which the fossils have been found gives us a chronology that dates back some seven million years to that time when we shared a common ancestor with the chimpanzees. The timing of when that common ancestor lived is derived from the so-called molecular clock and seems, so far, to be supported by the fossil record. The recent discovery in Chad of a very ape-like extinct species that has been called *Sahelanthropus tchadensis*, nicknamed Toumaï, is very close to that critical point of divergence between the chimpanzee and human lineages.

One of the most important of recent developments in the unfolding history of our species is the realisation that evolution can be driven by the interaction between internal genetic changes and external changes in the environment. Over the last century scientists have discovered that what was at first thought to be geological verification of the biblical Flood story is in fact evidence of repeated dramatic changes in climate over the last couple of million years and more, and which we now recognise as the Quaternary ice ages.

The impact of climate change

It was against this background of a global deterioration in climate, from around five million years ago, that environmental change took place, even in the tropics. And it was these climate changes that forced changes in vegetation cover and thus the plant-eating animals, including our remote ape-like ancestors. Repercussions were felt up the food chain to the top predators, that by around two million years ago included our more recent ancestors who were themselves predators on plant eating big game, from elephant-sized animals downwards.

At present, the 7 million year old *Sahelanthropus* from Chad in Africa is our oldest known human relative subsequent to our evolutionary divergence from the chimps.

Only in the 1840s was it realised that much of Northwest Europe and North America had once been glaciated by an Ice Age rather than inundated by the Biblical Flood.

In the high latitudes of the northern hemisphere the often rapid oscillations in climate had drastic effects on the survival of our extinct relatives who first ventured beyond the tropical climates of Africa. It is still not entirely clear how and to what extent climate did impact upon them but there is growing evidence that climate change was a major force in shaping the movements of human relatives and ancestors especially over the last few hundred thousand years.

Mendel's all-important contribution

The other major scientific revolution that has impacted upon our story is that of genetics. Darwin did not have an understanding of even the basics of genetics available to him to underpin his theory of evolution. Although initial evidence was gathered and even published by the Austrian monk, Gregor Mendel (1822-84), in 1865, only seven years after the Darwin-Wallace outline theory of evolution, Mendel's all important data, analysis and insights were not widely noticed or appreciated. Not until 1900 were the Mendelian laws of inheritance fully recognised. And then it was another 53 years before the key structure for inheritance was discovered – the double helix of deoxyribonucleic acid (DNA). Francis Crick (1916-2004) and James Watson (1928-) worked that out in the early 1950s. They showed how the molecule could be replicated by the unravelling of the double helix strands, each of

James Watson and Francis Crick's discovery, in the early 1950s, of the double helix structure of DNA provided a key to understanding the mechanism of inheritance through duplication of the genetic code.

which would carry a copy of information encoded by a sequence of the enzymes represented by the letters ACGT.

It is a digital code but it is not binary. It is quaternary, with four distinct items. The encoding information is contained in an ordered sequence of four different symbols called 'bases', namely Adenosine, Cytosine, Guanine and Thymine. Hence the application of the A, C, G, and T designation.

These four substances are the fundamental 'bits' of information in the genetic code, and are called 'base pairs' because there are actually two substances per 'bit'. Everything else is built on top of this basis of four DNA digits.

Mapping the human genome

So, a vital key had been found, but even then it took another 50 years for adequate techniques and understanding to be developed before we have been able to map the human genome. Now that is virtually complete we can really begin to read the 'encrypted' biomolecular story of our development and our relationships to other organisms. Along the way important insights have been gained into the way we see ourselves.

Our very close genetic relationship with one another can only have survived because we are a recently evolved species. Inevitably, there are numerous small genetic differences between living people that have accumulated over the last 100,000 years, but the vast majority of them are to be found within populations and only a minority represent differences that have built up between populations. None of them are significant enough to prevent us from interbreeding.

DNA – the fragile key

The biomolecular evidence not only emphasises our human commonality and recency as a species but also provides strong support for our ancestry in Africa. But we are far from knowing the whole story. Unfortunately there are severe limitations on the recovery of ancient biomolecules from the fossil record. DNA is a particularly fragile molecule and even in the most favourable of post-mortem environments – ice or dark, cool and dry caves – only fragments can be recovered and amplified. So far the only ancient DNA to be recovered so far

from our extinct human relatives has come from some Neanderthal bones found in high latitude cave environments. However, the hope of recovering DNA from ancient relatives and ancestors that lived in the tropics is remote. Even cave dwelling species like the recently discovered diminutive *Homo floresiensis* people of the Indonesian island of Flores are unlikely to have any of their DNA preserved in the remaining bones.

For the immediate future the mapping of the chimpanzee genome will provide fascinating insights into what makes the difference between being a chimpanzee and being a human, but it is just as likely to raise as many questions as it solves. Then the possibility of mapping the gorilla genome will become more urgent because it will put some of the chimpanzee genomic differences in an evolutionary perspective.

Meanwhile the hunt for more human-related bones and stones will continue and undoubtedly throw up a host of new questions and answers. Remember that most of the 20 or so extinct relatives that we know about have been discovered in the last 50 years. Many of them are still only known from a few bones. There are very few complete skeletons and overall our fossil record is very fragmentary and incomplete. There is plenty of scope for future finds of great significance to be made by anyone with enough determination, patience, skill and luck.

A computer display of the DNA sequence of bases that make up a human gene. The sequence is shown both graphically and labelled (coded) with the initial letters of the relevant four bases that make up DNA ie guanine (G), cytosine (C), thymine (T) and adenine (A).

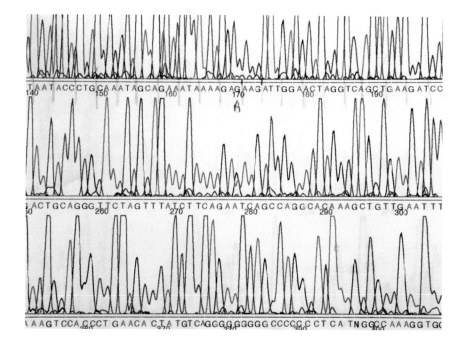

From football fans to native Papuans, face paints and decoration is still used by celebrants of games and ritual ceremonies. Powdered red ochre has been used for such purposes by modern humans for at least 70,000 years and seems to be a specific behavioural attribute of *Homo sapiens*.

PART I

This section sets the scene for our investigation of ourselves as a biological species with an evolutionary history extending back over six million years. What kind of evidence is available, how good is it and how can we interpret the often puzzling information? The way that we see ourselves has been revolutionized over the last few centuries. From being one rung below the angels we are now almost 'kissing cousins' to the chimps.

CHAPTER ONE

Humankind –
a truly global species

Biologically we humans are an interesting species. We are also enormously successful (so far) because we have a booming and globally distributed population that is very well adapted to a very wide range of life styles, environments and habitats. But we are also unique among all the life forms on earth, as far as we know them, for we are self-reflective... and what makes us even more unique among life forms is that we have a concept of death and an afterlife. We have developed many different techniques for investigating ourselves, our relationships with one another, with other life forms and the inorganic world around us.

Chimpanzees (*Pan troglodytes*), like humans, use tools to obtain food. Here a chimpanzee in captivity is using a stick as a tool; in the wild chimpanzees 'fish' for termites with a stick. Chimpanzees also use leaves to carry water and stones to crack open nuts.

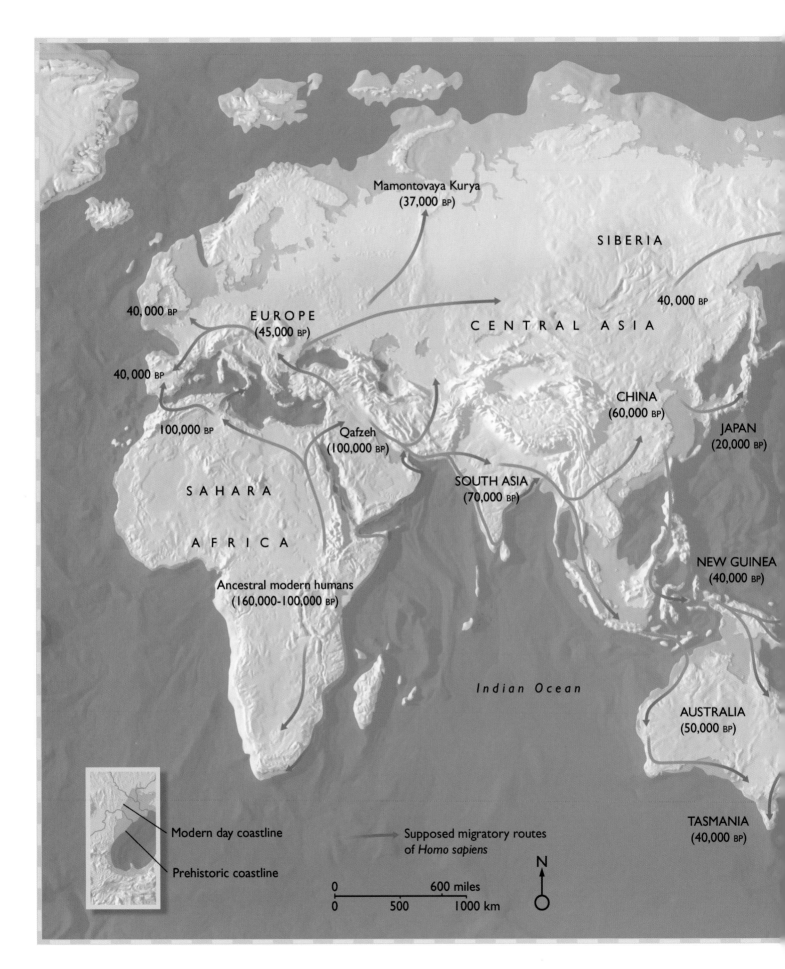

Mamontovaya Kurya
(37,000 BP)

SIBERIA

40,000 BP

EUROPE
(45,000 BP)

CENTRAL ASIA

40,000 BP

40,000 BP

CHINA
(60,000 BP)

JAPAN
(20,000 BP)

100,000 BP

Qafzeh
(100,000 BP)

SOUTH ASIA
(70,000 BP)

SAHARA

AFRICA

NEW GUINEA
(40,000 BP)

Ancestral modern humans
(160,000–100,000 BP)

Indian Ocean

AUSTRALIA
(50,000 BP)

TASMANIA
(40,000 BP)

Modern day coastline

Prehistoric coastline

Supposed migratory routes
of Homo sapiens

N

0   600 miles
0   500   1000 km

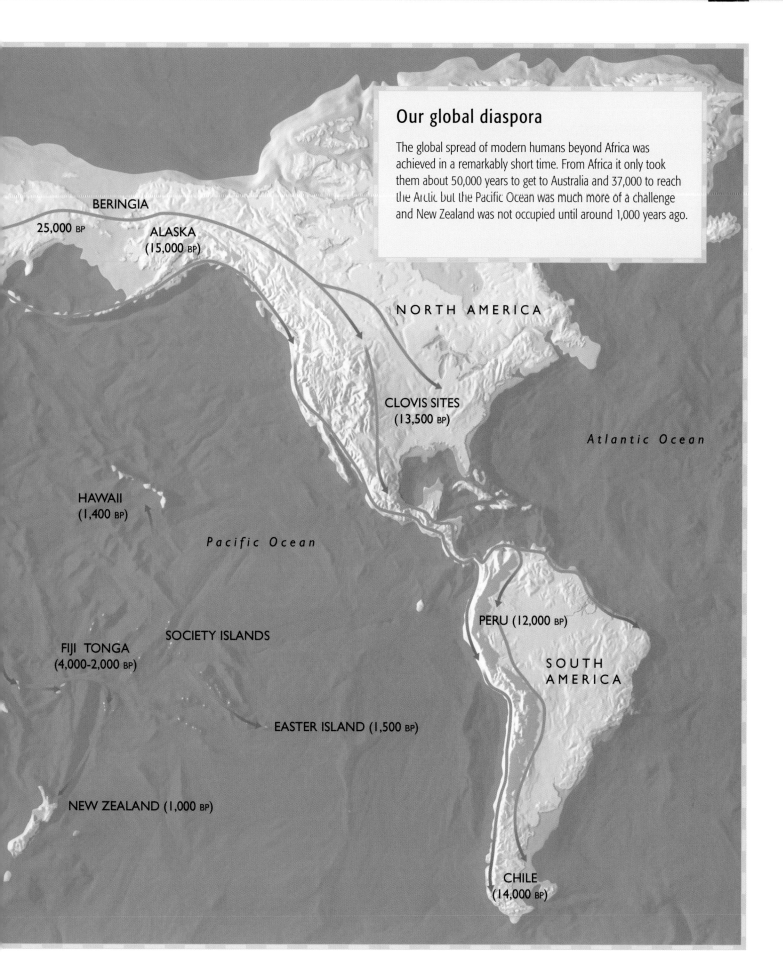

Our global diaspora

The global spread of modern humans beyond Africa was achieved in a remarkably short time. From Africa it only took them about 50,000 years to get to Australia and 37,000 to reach the Arctic but the Pacific Ocean was much more of a challenge and New Zealand was not occupied until around 1,000 years ago.

BERINGIA

25,000 BP

ALASKA
(15,000 BP)

NORTH AMERICA

CLOVIS SITES
(13,500 BP)

Atlantic Ocean

HAWAII
(1,400 BP)

Pacific Ocean

SOCIETY ISLANDS

FIJI TONGA
(4,000-2,000 BP)

PERU (12,000 BP)

SOUTH
AMERICA

EASTER ISLAND (1,500 BP)

NEW ZEALAND (1,000 BP)

CHILE
(14,000 BP)

When Carl Linnaeus first formalised our scientific name in the tenth edition of his *Systema naturae*, published in 1758, he chose the appellation *Homo sapiens*, meaning 'knowledgeable man'. He also listed four 'varieties' *Europaeus 'albus'*; *Americanus 'rufescens'*; *Asiaticus 'fuscus'* and *Africanus 'niger'*. As we shall see, this crude categorisation on the basis of skin colour was typical of this historical period when it was still not clear how living humans related to one another.

As we now know we are all one interbreeding species and Linnaeus's varieties are not significant at the species level. For a description of our species, Linnaeus suggested that we 'know ourselves' (*nosce te ipsum*). For thousands of years humans have expended much time and effort trying to do just that by inventing a whole panoply of mental and physical methods and disciplines for our investigations. Today we categorise them into recognisable subject areas such as philosophy, psychology and numerous compartments of science of which biology and anthropology are our main concern here.

Most populations and many species go through 'boom-and-bust cycles', the fossil record is littered with them. Is there any reason why we should be exempt?

Populating the Earth

Overall, we are undoubtedly a very successful species, although so far we are a relatively short-lived one in evolutionary terms. Over the last 100,000 years or so we have grown from a total population of a few hundred thousand to a global population that is close to 6.5 billion – and still rising.

Within this 100,000-year history, however, the global population has only really exploded with exponential growth over the last few tens of thousands of years. Before that, it grew relatively slowly because the population was still dispersing and dependent upon a mobile hunter-gatherer mode of life. The real boom did not get under way until there was more permanent settlement 'fuelled' by the early development of agriculture on some of the most fertile river floodplains of the Old World such as the Tigris and Euphrates in Mesopotamia, the Nile in Egypt, the Indus and Ganges in India and the Yuang Ho in China.

Hopefully, our growth rate will begin to tail off now as many nations grasp the 'nettle' of self-

Our close relationship with the great apes has been recognised for over 300 years. In 1758, the Swedish taxonomist, Carl Linnaeus, was the first to formally classify our species as *Homo sapiens* along with the great apes within a larger group which he called the Primates.

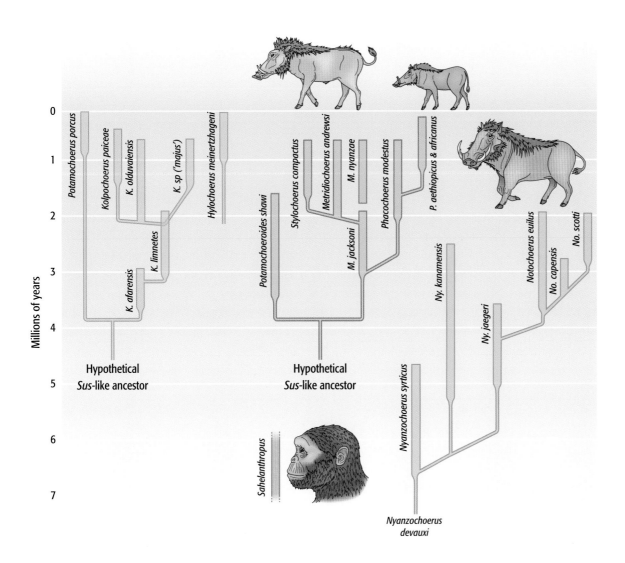

Wild pigs (suids) have been common animals in Africa and have left a good fossil record of their rapid evolution which has now been calibrated against chronological time. Suid fossils have proved invaluable in the relative dating of human fossils, especially where local radiometric dates are not available as in the case of *Sahelanthropus* whose 6-7 million year old age has been determined by correlation with suid fossils.

regulation in terms of birth control. Otherwise our 'boom' is likely to be followed by a catastrophic 'bust' as we continue to stretch our food, water and energy resources to the limit. The same cycle of 'boom and bust' has happened to many other species over the past 3.5 billion years or more during which life has 'graced' planet Earth and we have to remember that the vast majority of life forms that have existed on Earth are now extinct.

Is there any reason why we should be exempt?

Consider this: the United Nations estimates the global population will reach 9.1 billion by 2050. Our current birth-rate (2005) is just over 3.9 babies *per second*, and our current global death rate from all causes is just over 1.7 *per second* – so we're gaining something like 2.2 new mouths to feed, *every second*.

Compared with a multitude of other species, including many of our extinct relatives, our existence as a species so far has been very brief. Many

mammal species survived for up to a million years or so before becoming extinct or evolving into one or more 'daughter' species. Our understanding of our nearest extinct relatives who have lived over the last six million or more years is still very fragmentary and several species are known from nothing more than a handful of fossil bones. Some of the best-known species lasted for anywhere between a few hundred thousand years (the Neanderthals, for example) to over a million years (*Homo erectus*). However, all these species had much smaller total populations than we have.

At present we are so numerous that it would be difficult for us to become extinct within a very short span of time. Even in the event of a real catastrophe, such as a large asteroid impact from outer space – the kind of event that helped terminate the dinosaurs – or a global pandemic disease, there would almost always be residual surviving populations in more protected or isolated environments such as remote oceanic islands.

Then there would be the question, however, of whether the survivors would be numerous enough to recover and repopulate the earth to any significant extent.

It is likely that the now-extinct Neanderthals faced a similar population-threatening scenario around 35,000 years ago. At that time, during the last glacial of the Pleistocene ice ages, average annual temperatures in central France plummeted towards zero and the Neanderthals were pushed back from their homelands in Northern Europe to refuges in the far south near the Mediterranean. During this time, tundra grasslands extended right down to Greece and Spain.

Although the climate became milder within a few million years, the Neanderthal population was probably so small and fragmented that they were never able to recover, and by 28,000 years ago they were extinct. By comparison, modern humans – who had also been pushed southwards by the glacial conditions – were able to recover because they had a much bigger population pool from which to draw, and perhaps more advanced social and cultural assets that helped recovery.

What characteristics do we have that allowed globalization?

The spread of humankind

One of the most remarkable aspects of 'our' human story is the history of our globalisation as a species, which happened relatively quickly but not a great deal faster than that of other animal and plant migrants in recent evolutionary history.

The fact is, much of life is highly opportunistic and will expand in numbers and radiate geographically within the tolerance limits of the species. In short, if the conditions are right for a particular organism, it will take advantage of them to the absolute limit, to grow and multiply as far as it is able to do so.

Around 120,000 years ago early members of our species who were adapted to African climates spread both within and without the vast continent of Africa. By 60,000 years ago our ancestors had reached Australasia, and the Americas by around 20,000 years ago. The last great environment left to conquer was the island-studded expanse of the Pacific Ocean, which covers about a fifth of the Earth's surface. To successfully traverse this ocean demands great skills of seamanship and navigation; these were eventually developed and populating the Pacific was finally achieved by around 1,000 years ago.

The only remaining major continent and environment left to conquer was Antarctica. However, with its frigid polar climate it was not 'colonised' until the 20th century AD – and then thanks only to modern technologies in the fields of clothing, shelter, energy and food supply.

Our biological travelling kit

It is worth asking what characteristics our species have which allowed such an expansion.

The basic elements of our biological travelling kit include a backbone and four limbs. Our backbone of interconnected vertebrae is, by definition, a common characteristic of all vertebrates. As a group, the vertebrates have been enormously diverse and successful from Carboniferous times over 300 million years ago, although they originated as far back as early Cambrian times some 540 million years ago. Throughout this long evolutionary history they have diversified from fish and amphibians to dinosaurs, birds and whales. Even so, the invertebrate arthropods, especially the insects, are still more numerous, diverse and have a greater biomass than the vertebrates. In short, all of the Earth's invertebrate arthropods and insects in total weigh more than all of the Earth's vertebrates.

Viruses spread like 'wildfire'

Viruses can spread globally very rapidly by 'piggy-backing' on mobile vectors (carriers) such as migratory birds and modern human travellers. As an example of this ability to 'travel', the bubonic plague that struck Europe in 1348 springs to mind as a human disaster of the first magnitude.

In this case the disease, caused by the bacillus *Yersina pestis*, carried by the fleas on the black rats that dispersed throughout Europe, took barely three years to run its course... By around 1350 it was over – as were the lives of as many as 23 million of its victims. It is a classic example of how a disease-causing organism can spread with frightening speed.

Nowadays of course, AIDS, SARS, Asian bird flu and the pathogens causing other ailments demonstrate the same ability to spread when given the opportunity.

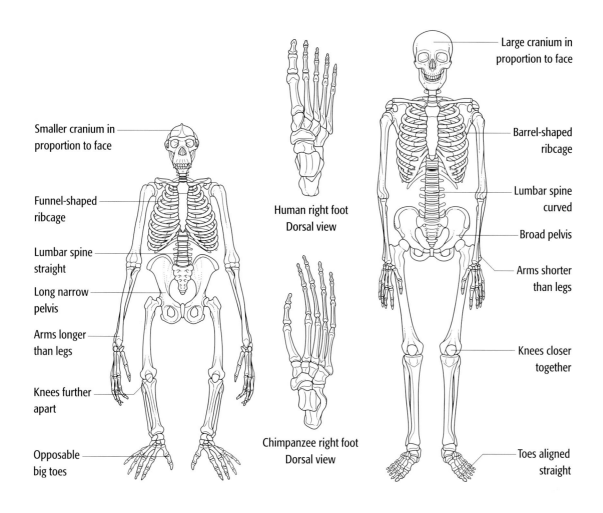

Smaller cranium in proportion to face

Funnel-shaped ribcage

Lumbar spine straight

Long narrow pelvis

Arms longer than legs

Knees further apart

Opposable big toes

Human right foot Dorsal view

Chimpanzee right foot Dorsal view

Large cranium in proportion to face

Barrel-shaped ribcage

Lumbar spine curved

Broad pelvis

Arms shorter than legs

Knees closer together

Toes aligned straight

Comparison of the skeletons of a chimp and human show how similar we are in terms of the basic bony framework but also how different in terms of relative proportions which are adaptations for different lifestyles. The chimps, with long powerful arms and long-toed feet are well adapted for climbing in trees whereas we are well adapted for walking upright. Even our rib cages reflect our very different eating habits, with the chimp ribcage accommodating the large belly of a plant eater whilst we have a waist and the small belly of flesh eaters.

The backbone protects the central nerve cord and provides the body with an essential flexible, elongated strengthening column to which numerous skeletal structures are hung and to which many groups of muscles are attached. The tetrapod or four-limbed vertebrates, which include amphibians, reptiles, birds and mammals, have two girdles – pectoral and pelvic – which provide attachment and articulation for the arms (or forelegs or wings) and legs (or hind legs) respectively.

The skull articulates with the front end of the backbone and its bony 'box' protects the main sense organs of sight, hearing and smell along with the swollen end of the central nerve cord that we recognise as the brain. Finally, the rib cage protects the lungs, heart and liver and other organs in the thorax.

In most tetrapods the backbone is held parallel to the ground surface or seabed with the limbs and body organs hanging down from it. However, in land-living tetrapods the limbs act as struts that

hold the body clear of the ground surface against gravity to reduce friction during locomotion. The limb struts as previously mentioned are attached to the backbone via the girdles, with the result that the backbone tends to sag between the two main struts like a suspension bridge. The head and tail act as counterbalances, particularly for those animals that have bulky guts.

The first mammals

The tetrapods we are most concerned with are the mammals that first evolved in a very primitive form in latest Triassic times (251 to 200 million years ago). By late Jurassic times (200 to 145 million years ago) the more advanced placental mammals had evolved and they were all small tetrapods about the size and shape of a shrew. Biologically the features that distinguished them from other tetrapods such as the amphibians and reptiles, were warm bloodedness with body hair for insulation and thermoregulation.

Like all primates we humans are mammals who give birth to live young. In addition our primate heritage has opted for a reproductive strategy in which a single baby, and rarely twin or more babies, are born at a relatively large body size but still in need of prolonged maternal breastfeeding, care and protection.

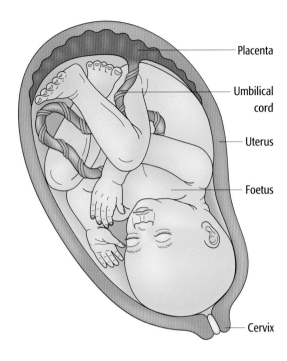

Placenta

Umbilical cord

Uterus

Foetus

Cervix

There are still some 230 species of primate, mostly but by no means all, living in tropical woodlands and forests. They range from 180kg male gorillas to 100g tarsiers.

For all mammals, internal fertilization leads to the development of the young within a uterus where they are connected with the mother via a placenta.

Following birth, the young are fed from milk produced by mammary glands. In addition the lower jaw is reduced to a single bone (the dentary) with the remaining bones forming the three small bony elements (ossicles) of the middle ear – hammer, anvil and stirrup (malleus, incus and stapes). Mammalian teeth have different forms for different functions – incisors, canines and molars. The incisors at the front are blade or chisel-shaped for cutting and holding, the dagger-shaped canines at the front corners are for stabbing and holding and the complex molars further back are for crushing and grinding.

All these characteristics mean that generally the hairy and warm-blooded mammals can be continuously active even at night or in the cold if necessary. But they need a constant supply of food to energise such activity. Cold-blooded reptilian crocodiles can survive for months without food and there are some snakes, such as pythons, that eat only once a year. However, warm-blooded mammals require 'fuel' far more frequently; lions, for instance, need a couple of kilograms of raw meat protein a day just to 'tick over'. To be active, mammals need a 'quick fix' from their food and that means releasing as much energy as quickly as possible by

preprocessing it before it even reaches the stomach. The teeth act as a food processor, chopping up and breaking down the animal or plant tissue before it is swallowed.

The mammal's internal development of the young via a placenta to the mother's energy supplies means reproduction need not be as wasteful as the amphibian or reptilian method of laying many eggs so that some may survive. Furthermore, breast-feeding promotes the rapid development of mammalian babies. Note that the warm-blooded birds have adapted a basically reptilian biology so that it is virtually as successful as that of the mammals. But within the 5,400 or so species of mammals we are primarily concerned with one small group, the primates (Order Primates) and within them the even smaller grouping of the hominoids (Superfamily Hominoidea).

Primates – the key features

Primates, hominoids, hominids and hominins include the monkeys and apes.

Today there are still some 230 primate species, mostly but by no means all living in tropical woodlands and forests. They range from 180kg male gorillas to 100g tarsiers and include all the apes, monkeys and prosimians (such as lemurs and bushbabies). Key features of the primates are forward facing eyes which provide binocular vision and good depth perception; nails instead of claws; five fingers and toes including an opposable thumb on the hand; a long heel; relatively large brain and long childhood. Most of these characteristics are adaptations for tree dwelling inherited from small insect eating and tree-dwelling ancestors that lived over 70 million years ago.

The hominoids include the gibbons (also known as the lesser apes), higher apes (chimpanzees, gorillas and orang-utans) and us along with our fossil ancestors and relatives. As we shall see in Chapter Three, in the 19th century there was a debate as to which of the ape groups is closest to humans. Darwin selected the African higher apes while Haeckel chose the gibbons of Southeast Asia and we now know that Darwin was right. It is worth briefly considering what features unite the hominoids as a group. The apes are tailless and generally the arms are longer than the legs because they are used in moving around, especially climbing and arm swinging (brachiation) through woodland and forest canopies. The gibbons with their long arms are especially well adapted for brachiation; in addition this type of locomotion is aided by their flexible wrist structures. Leaving the gibbons aside, the remaining grouping of the great apes of Africa (chimpanzees and gorillas) and Southeast Asia (the orang-utan) along with humans and our extinct relatives, form the Family Hominidae. For our purposes, however, we can be even more exclusive and focus just on the African apes and humans who together form the Subfamily Homininae.

The African apes are distinctive in their use of knuckle-walking on the ground, and, with regard to their feeding habits, the apes are essentially plant eaters – although they occasionally supplement their vegetarian diet with animal protein. This ranges from termites and ants (using tools to capture them) to raw meat from other mammals. The latter is obtained by male hunting parties that work together to capture and kill small monkeys, deer and pigs when available, and the whole exercise performs a variety of functions – dietary, social and sexual. The meat is shared among other members of the group and used to obtain sexual favours from the females.

So, the hominins include the chimpanzees, gorillas and our extinct fossil relatives plus us. And to recap on our biological heritage, we are all warm-blooded, hirsute tetrapods whose primitive stance is on all fours, although all can rise up onto the two hind limbs when necessary and thus free the hands for other purposes. Plant eating was the primitive mode of feeding which requires a lot of time spent finding, gathering and consuming this type of essentially low-energy food.

To derive maximum benefit, preprocessing by mastication is essential and then the plant material has to be stored in a capacious stomach for the gut microflora to aid digestion and release the food energy. As a result of the large stomach, the rib cage flares out at the bottom and is narrower at the

Comparison of the face and lower jaw of a chimp, australopithecine and human shows that the australopithecine face is much more similar to that of the chimp than the human. However, there is a greater similarity in the size and form of the human and australopithecine teeth, indicating less reliance on plant food and the use of teeth as weapons.

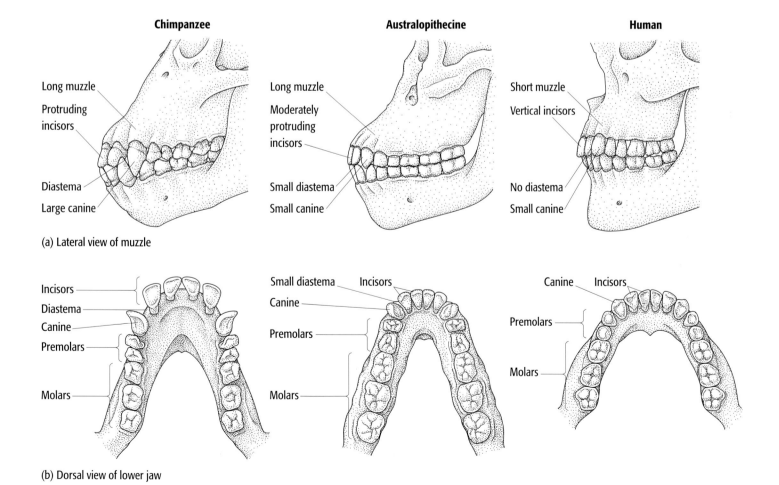

Chimpanzee **Australopithecine** **Human**

(a) Lateral view of muzzle

Chimpanzee: Long muzzle, Protruding incisors, Diastema, Large canine

Australopithecine: Long muzzle, Moderately protruding incisors, Small diastema, Small canine

Human: Short muzzle, Vertical incisors, No diastema, Small canine

(b) Dorsal view of lower jaw

Chimpanzee: Incisors, Diastema, Canine, Premolars, Molars

Australopithecine: Small diastema, Incisors, Canine, Premolars, Molars

Human: Canine, Incisors, Premolars, Molars

shoulders. There are significant differences in size and strength between the sexes (sexual dimorphism), which are greatest in the gorillas (and in the hominoid orang-utan,) where males may be twice the weight of the females. Apart from sheer strength, the males use their primitively prominent canine fangs as effective weapons and they can inflict terrible wounds with them. Indeed they are used to killing when hunting.

The role of sex

Both the chimpanzees and gorillas retain the primitive mode of reproduction whereby the females are only fertile at intervals and there is male competition for dominance and sexual access to the females. However, the situation is very different in the case of the misnamed pygmy chimpanzee species (*Pan paniscus*) more commonly known as the bonobo, which is actually no smaller than its close relative, the common chimpanzee (*Pan troglodytes*).

The bonobos use sex widely for social purposes. Social groupings are quite well defined but differ in structure between the chimpanzees and gorillas, but both are territorial and are constantly vigilant regarding their group boundaries. Most disputes are settled with posturing and shows of strength but mortal conflicts do occur.

Reproductive biology is very similar between the great apes and humans with just one or two offspring resulting from any one pregnancy. The babies are pretty helpless at birth and require prolonged breast-feeding and maternal care before they are able to become more independent. Lactation helps suppress ovulation in the mother so that they do not usually become sexually receptive and pregnant until the previous infant is old enough to be weaned. However, as noted above, this does not apply to the bonobos nor to humans.

Ape infants develop more quickly than human ones in the early stages of life, being very mobile almost from birth, but they are overtaken by human infants around the age of four. Adolescence in apes occurs around the age of eight or nine when the females become sexually receptive for the first time but they do not conceive for another two or three years. There are marked secondary sexual changes, especially in male apes, with bony eyebrow ridges becoming much more prominent, and in gorillas

The ape's sense of smell is so acute that males identify their own offspring through smell – despite playing little if any part in the young's upbringing. Their sight and hearing are similar to our own.

there is also a marked bony crest to the skull for attachment of the massive jaw muscles.

The senses of sight, smell and hearing in the apes are very similar to ours. They have good stereovision – essential for fine depth perception – and colour vision for knowing when favourite fruits are ripe. If anything, their sense of smell is considerably better than ours and males can identify their offspring even though they take little or no part in their upbringing. Indeed, newly dominant males will attempt to kill the babies of their predecessor and may intervene in disputes to help their own offspring. Their relatively large brains are perfectly adequate for their social way of life. In fact tests show that the cognitive and learning potential of the higher apes is much greater than they seem to need or use in the wild. But this may just be a reflection of our ignorance of their social interactions and thought processes.

Interaction and communication

Their energy-rich diet of fruit and young leaves means chimpanzees do not have to spend as much time each day eating as the gorillas do. Chimpanzees may eat from at least 20 plant species each day and have an overall choice of some 300 plant species, which they monitor continuously, so much of their time is spent travelling around their territory looking for ripe fruit and young leaf shoots. When they find it they eat as much as possible, rest for an hour or two, and then move on to the next possible food source. There is plenty of opportunity for social interactions within the group and at night they individually make nests of branches and leaves high in trees. The nests are never reused, however – perhaps to avoid parasites or predators.

Although chimpanzees have a good range of vocalisation with some 13 categories of calls being recognised, they have nothing like speech as we understand it, and cannot be taught to speak. This is because of the structure of their vocal cords and the coordination of the tongue and breathing that is necessary for some complex speech patterns. However, their intelligence is acute enough to learn sign language with a much bigger vocabulary than their range of vocalisation would indicate. They can also learn to operate computer-coded words and string together some simple commands, though not

Chimp development like that of humans is 'hardwired' for mimicry of parental behaviour which allows the acquisition of complex behaviour, such as tool use.

grammatical structures. A captive gorilla by the name of Koko is reported to understand 2,000 spoken words and 1,000 hand signs in American Sign Language (ASL). But then even trained orcas can pick up ASL and differentiate between some simple syntactical differences such as 'spit and go' and 'go and spit'.

Chimpanzees' level of self-awareness is the highest of all other animals apart from us and they are capable of limited 'mind reading' or pre-emptive behaviour based on precognition of what another is likely to do in a particular situation. There is still an on-going argument about the level of consciousness attained by the apes, especially the chimpanzees, but most tests have been carried out on captive chimpanzees that have been 'humanised', often to a considerable degree.

Our understanding of ape biology and behaviour has increased enormously within the last decade or so with many prolonged and intensive studies of the animals in the wild, but we have only been able to scratch the surface here. Until recently a major problem for primatologists has been the recognition of the real genetic relationship between various chimpanzees in the group being studied.

For instance, it used to be thought that adolescent male bands tended to be fraternities of blood-related half siblings and true 'brothers'. However with the advent of DNA testing using faeces it has been discovered that social relationships and bonding are much more flexible, with collaboration between unrelated 'friends'.

The best model?

So, which living relative is the best model for our extinct ape-like ancestors? Although genetically we are seemingly very close to the chimpanzees and share some 98 per cent of our genome with them, according to the molecular clock, that two per cent difference amounts to some six million years in real time since we last shared a common ancestor.

In evolutionary terms that is a long time, especially for a rapidly evolving genus like ours. Furthermore, we have virtually no information about the evolution of the chimpanzees or gorillas since that common ancestor, as there is hardly any fossil record (see box p30). The main reason for this is that fossils of animals that live in woodland and

Chimpanzees are capable of limited 'mind reading'. They can anticipate what another is likely to do in a particular situation, and then move to pre-empt it.

The missing fossils

One of the most problematic gaps in the fossil record is the lack of any fossil remains of the African apes. Until 2005 it has been true to say that there are none known but this is no longer so. Three teeth have been found in 500,000-year-old sediments of the semi-arid Tugen Hills in the Kenyan stretch of the Great East African Rift Valley.

American anthropologists Sally McBrearty and Nina Jablonski have described the fossils and their location, which lies 600km (375 miles) to the east of the present and historic range of the chimpanzees that are confined to wooded regions of central and western equatorial Africa. The Tugen Hills site has also yielded fossils of our more direct *Homo* relatives who were therefore neighbours and contemporaries of these chimpanzee ancestors.

At the time the site was wooded with lakes, streams, active vulcanicity and abundant game that ranged from fish, turtles and crocodiles, through rodents, hippos, wild pigs and cattle to elephant and a species of colobine monkey.

Although the chimp fossil record still only consists of three teeth that do not tell us very much about chimp evolution, their preservation indicates the potential for further finds of chimpanzee fossils in the region.

Comparison of two kinds of extinct human relatives shows how extensive evolutionary change has been within a few million years. The little *Australopithecus afarensis* ('Lucy' and her kin) is very ape-like despite its bipedal stance. By contrast, the skeleton of *Homo ergaster* is very like that of a modern human, but its brain size was about a third of ours.

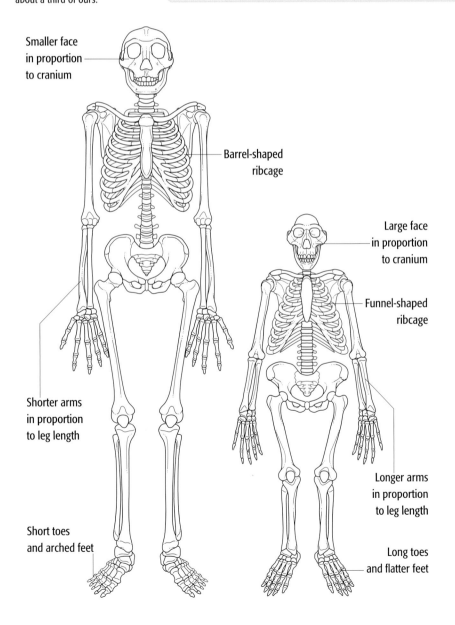

Smaller face in proportion to cranium

Barrel-shaped ribcage

Large face in proportion to cranium

Funnel-shaped ribcage

Shorter arms in proportion to leg length

Longer arms in proportion to leg length

Short toes and arched feet

Long toes and flatter feet

forest environments are very rarely preserved in sediment as the soils in these environments are far too biologically active. The result is that both invertebrate and microbial activity soon breaks down most organic materials.

We do not know the extent to which the chimpanzees and gorillas have evolved over the last six million years although there are reasons to believe that they might have been pretty conservative since they are well adapted to their habitats and these were much more extensive in the past.

However, some experts think that the chimpanzees may actually be quite highly adapted and may even have descended from a bipedal ancestor. Mapping of the chimpanzee genome, which has recently been outlined, will undoubtedly resolve a lot of these questions.

By contrast, our human-related evolution has not only been rapid but also has radiated into a 'bushy' shaped family tree with many branches that have become extinct. Hence, ape models for the way our ancestors and relatives might have lived and behaved, if they apply at all, have to be treated with caution. All the evidence points to upright (bipedal) walking being achieved by our extinct relatives at least five million years ago and perhaps well before that, but the evidence is still not unequivocal.

There is no doubt that by the time the earliest australopithecines evolved over four million years ago they could walk bipedally, even if not in a completely modern manner, and they were still fairly adept at climbing trees, again if not as well as their true ape ancestors – so it was a typical evolutionary compromise. Evidently, the benefits of walking upright outweighed the dangers of being less adept at tree climbing, presumably freeing the arms and hands for other tasks. We also know that these small primates were still quite ape-like in many aspects, with a brain that was still chimpanzee-sized.

There were a number of other downsides to this peculiar bipedal stance for a fundamentally quadrupedal mammal. The backbone and lower limbs had to bear all the weight of the body and this was exacerbated by walking when the weight of the body was 'hammered' by alternate heel impacts up through the legs, pelvis and backbone to the base of a heavy skull. To cope with the stresses the backbone had to be able to absorb the compressive shocks and this was achieved by means of its double curvature and the intervertebral cartilaginous discs. The stresses on the legs were reduced by changes in the foot bones, heel and bent knee that can also be locked straight when necessary. The legs themselves had to be brought in under the body with the upper leg bones sloping inwards and slight changes in the articulation with the pelvis. Chimpanzees and human infants waddle when they walk because their legs are set further apart and so the body has to be swung around with each step to maintain balance.

The pros and cons of a bipedal stance

These changes in the pelvic area have a downside when it comes to reproduction in bipedal human relatives as they make the birth process much more laborious and potentially dangerous for both mother and infant.

However, the upright stance allows mothers to carry infants at the breast and even suckle them while engaged in other tasks such as gathering food. With the female genitalia being largely obscured from view by the vertical body position, the possibility of signalling fertility and sexual receptiveness by enlargement or colouration of the genitalia as seen in the higher apes is no longer possible. Presumably this is when a greater emphasis was placed on the sexualisation of behaviour and nonverbal communication.

Exposure of an upright body to the tropical sun also means that the top of the head receives much more direct heat than does that of the knuckle-walking and woodland dwelling apes. The brain is a delicate organ that can easily become overheated, leading to permanent damage or even death, so the adaptation of protective head hair becomes distinctly advantageous. However, simultaneously the insulatory effect of ape-like body hair is less advantageous.

Sweating is an efficient means of cooling the body provided the water loss can be easily replenished, but that is a tall order in semi-arid and arid tropical regions. The alternative is to lose the body hair and protect the skin from the damaging effects of the tropical sun by using the natural pigmentation already present in the body.

Skin colour – an evolutionary compromise

Historically, skin colour has been used as an excuse for some of the most extreme racism and is still a topic prone to all kinds of prejudice in some parts of the world. Discovery and description of the global variation in skin colour by predominantly white European explorers was part and parcel of our rediscovery of ourselves as a global phenomenon which has its roots in classical times. The use of skin colour and other 'racial' or ethnic characteristics was intrinsically linked to the need to see other ethnic groups as inferior in order to justify enslavement. At times science was enrolled to help put a rational gloss on the whole sorry business.

For instance, the discovery of albinism (a genetic condition in which pigment is not expressed in the skin, hair or iris of the eye) in otherwise dark-skinned peoples, was falsely claimed to indicate that white skin was natural and that darkened skin was therefore somehow degenerate.

Skin colour, climate and the sun

Exposing ourselves to sunlight, as we all know, brings both benefits and risks. Sunlight is essential for the synthesis of Vitamin D that is needed for calcium fixation and bone growth, but it also destroys Vitamin B folate through photolysis, and deficiency in this member of the Vitamin B complex can cause anaemia. Continuous exposure to sunlight also promotes skin cancer.

Albinism in otherwise dark-skinned peoples was falsely claimed to indicate that white skin was natural and that darkened skin was therefore somehow degenerate.

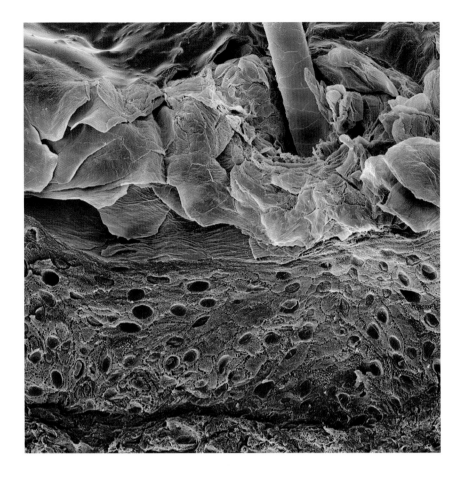

Human skin contains melanocytes (coloured dark brown in this section) that produce the skin pigment melanin. This darkens (tans) the skin when exposed to strong sunlight and helps protect it from damaging ultraviolet light that can cause skin cancer.

Sunlight in itself is not the problem, ultraviolet radiation is the main culprit and its magnitude, and hence the danger it poses, is determined by a number of factors.

The activation of pigmentation in the skin is a potential we all have and it helps protect the skin against damage, but at the same time can block the benefits. For example, black-skinned people living in Scandinavia, especially those who have been there since childhood, risk serious Vitamin D deficiency unless they take vitamin supplements. Conversely, white-skinned people living in Australasia are prone to skin cancer and Vitamin B folate deficiency.

It has been thought that there should be a simple correlation between skin colour and latitude, and there is indeed a rough correlation, but there are many exceptions to be found in the distribution of modern humans. A complicating factor is this: it is not sunlight *per se* that is the problem.

Ultraviolet radiation (UVR) does the real damage. Its distribution and intensity are affected by a number of other factors apart from latitude, such as altitude, cloud cover and humidity. Mapping and correlation of UVR to skin colour explain over 70 per cent of the distribution of skin colour among modern human ethnic groups, which is a strong positive correlation, and some of the notable exceptions are readily explicable.

For instance five of the 12 most negative correlations, where skins are unexpectedly dark for the latitude, are found in Bantu-speaking peoples in southern Africa the populations of which are known to have migrated from equatorial regions to southern latitudes within the last few thousand years. Another of the negative correlation identifies the Inuit people of Greenland. They have unusually dark skin for the latitude in which they live but were able to get a natural Vitamin D supplement from a diet rich in sea-mammal protein. Today, however, many Inuit children suffer from a severe Vitamin D deficiency due to the modernisation and westernisation of their traditional diet.

Of the cases where peoples have unusually pale skins for their geographical location, three out the nine are correlated with peoples from Southeast Asia (Cambodia, Vietnam and the Philippines) who again have migrated, but this time towards the equator from high latitudes in recent millennia.

From this kind of evidence it has been argued that the extinct Neanderthal people who occupied high Eurasian latitudes for some 300,000 years would have become white-skinned, while the first modern humans who entered their territory 'hot foot' from Africa are likely to have been dark-skinned. Indeed the African ancestors of all modern humans would have been dark-skinned. However, it is not clear how long it takes for strong pigmentation to be lost.

The aboriginal people of Tasmania who were effectively wiped out by European settlers and their diseases are a tragic example. Even though they occupied the island of Tasmania for over 10,000 years their skin pigmentation showed no sign of getting paler. Tasmania lies around latitude 42 degrees south, similar to northern Patagonia in South America and New York and Barcelona in the northern hemisphere.

There are a number of other complicating factors, for instance sexual selection. In almost all peoples the women are measurably paler skinned than the men and that tends to ensure the positive benefits to their babies in terms of bone development. But this increased sign of fitness for motherhood may have become linked to increased desirability as a mate. Indeed skin condition is generally as good a sign of fitness in humans as

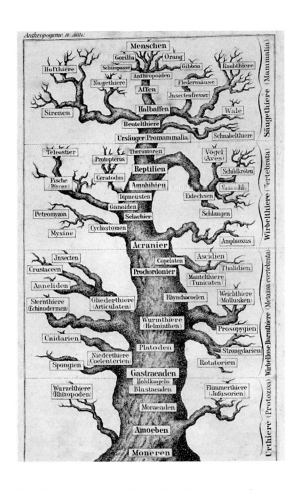

The German evolutionist Ernst Haeckel popularised the branching tree as a visual metaphor for the evolution of life and its classification with single celled organisms at the base and 'Man' (Menschen) at the top.

'healthy' plumage is in birds. Furthermore, male humans have traditionally used skin decoration as our equivalent of the coloured plumage of male birds as a sexual attractor.

It is only in the last couple of hundred years that men have to some extent renounced this habit except when signalling allegiance to a special cause.

Interestingly, both chimpanzees and gorillas have pale skins below their dark hair but dark skin on their faces where there is no hair and although chimpanzee babies have pink faces when born, the skin darkens as they develop. By contrast the shaded woodland-dwelling and more ancient orang-utans have relatively pale skins and red hair.

Bipedal 'apemen' inherit the Earth?

By the end of Pliocene times, 1.8 million years ago, there were at least three different upright walking, small-sized, small-brained, predominantly plant-eating and ape-like human-related species living in Central Africa. They were not particularly numerous, neither were they strong, capable of fast movement nor well armed with natural anatomical defences such as sharp teeth.

So it is likely that they were very vulnerable to predation and were regularly hunted down as food by a variety of predators including big cats, hyenas and big raptors. For instance, the Taung infant *Australopithecus africanus* (see p.137) has deep raking scratches on the skull and was almost certainly the victim of attack by a large eagle. Such raptors use their talons to kill prey and carry them bodily away. Other fossil skulls have been found with paired holes, which fit the gape of sabre-toothed cats. But from these vulnerable little 'apemen' evolved a hugely successful genus and series of species... but how and why did this occur?

One of the original arguments for our success, as promoted by Raymond Dart, was that our ancestors and relatives were in fact highly aggressive killer apes. This notion was brought into the general imagination notably by the hugely successful popular writings of American playwright Robert Ardrey, such as his 1966 book *The Territorial Imperative*. But when it was realised that the australopithecines were predominantly plant-eaters, this idea faded away fairly quickly. However, we do also know that almost right from the start members of our genus *Homo* had certain anatomical and behavioural differences from the earlier australopithecines. By around 2.6 million years ago the earliest representatives of our genus

Building and running brains requires a lot of energy, and that is expressed as the heat radiating from our heads.

Fossils of small ape-like human relatives and their young, like Raymond Dart's famous Taung skull, preserve teeth and claw marks showing that they were preyed upon, especially by big cats and raptors.

were already showing an increase in both brain and body size – but how and why was this happening?

The first thing we need to appreciate is that building and running a brain is very expensive in terms of energy, expressed as the heat radiating from our heads. Secondly, a larger brain and a big enough skull size to contain it create problems at birth, as any mother well knows, and a newborn infant's head is heavy and difficult to control. If such increases were to happen to a human relative who was still a plant-eater there would be the additional problems of a heavy, bony jaw and the musculature required to operate it.

The musculature would have to be modified to a considerable degree to allow inflation of the skull and the sheets of temporal muscles would become extended and inefficient. In addition, a great deal of time would have to be expended gathering and consuming plant food to fuel the body and brain. Thus a scenario incorporating such a large-brained plant-eater is improbable at this stage in human evolution.

Bipedalism had robbed the australopithecines of ape-style climbing skills and it would not have been easy for them to live purely off tender and succulent fruits from high in the tree canopy. So what fuelled the development of bigger brains?

We do know that tooth and jaw size was diminishing in some species such as *Australopithecus afarensis* even before brains became highly 'inflated'. We also know that primitive stone tools were being manufactured by 2.6 million years ago, although we are not entirely sure which of our ancient relatives first developed the technology. Furthermore, although the gathering of nutritious plant roots and tubers does not require climbing skills, these plant materials are tough and require considerable mastication – and that demands strong jaws.

More nutritious still is meat protein and there are two main ways to obtain it: scavenging and hunting. Most predators are not averse to opportunistic scavenging, driving smaller or fewer predators off their own kill, or feeding off the remains of prey left by the original predator. Scavengers include animals such as vultures and hyenas, though the latter are, however, fully capable of actively hunting down their own prey when they have to – just as the cats do.

We know that more advanced human-related species such as *Homo erectus* were proficient hunters, but it is likely that such behaviour initially developed from scavenging. Hyenas often deprive much bigger and more powerful hunters, such as lions, of their prey by harassment and sheer weight of numbers and then use their powerful jaws to dismember a carcass and remove the pieces to some place of greater safety for consumption. However, lions can also turn the tables on hyenas, who are also efficient predators in their own right. If there are enough lions they gang-up together to steal the hyenas' prey. As we have seen, muscular jaws were not a feature of the early members of our genus, but stone tools are just as effective at dismembering carcasses and smashing open long limb bones to obtain highly nutritious marrow.

Strength in numbers

The present model sees the possibility that even some australopithecines operated in groups to firstly spot kills made by other predators, often given away by circling 'eagle-eyed' raptors. Then they would have used their numbers and perhaps tools or weapons to drive off the original predator, dismember the cadaver and take what they could to some more protected site where they could consume their spoils in relative safety. The consumption of such high-protein food 'fuelled' brain development and that was advantageous because it provided the potential for greater cognitive skills. Furthermore, the necessity for collaborative group activity promoted certain social skills that in turn provided a positive feedback to the whole enterprise.

So though individually small and relatively weak, the situation changed when they formed active and increasingly intelligent groups of socialised scavengers. They achieved strength in numbers and by cooperating together as a team could achieve objectives beyond the reach of individuals.

By around 1.5 million years ago and the emergence of African *Homo erectus/ergaster*, an increasingly meat-based diet meant that there was less need for a capacious stomach and gut but a greater need for lung capacity, with the result that the rib cage changed shape. From being roughly funnel-shaped in the more ape-like plant-eaters it became more barrel-shaped as in modern humans

Out of Africa... the people who initially populated Africa then travelled to new worlds. They were the first of the world's great adventurers.

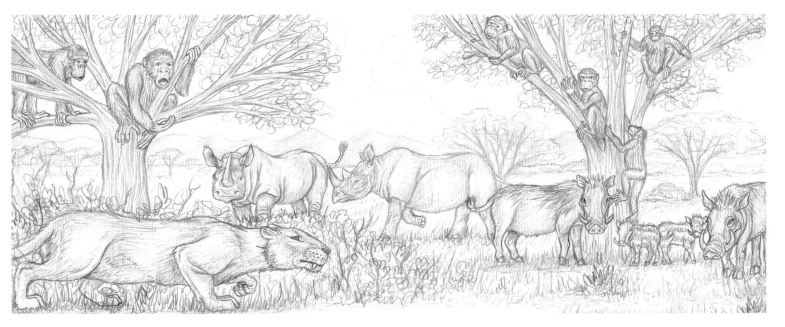

and consequently a waist developed below the rib cage and above the pelvis.

Legs became longer and arms proportionally shorter as the whole body became increasingly elongated and tubular to increase the surface-area-to-volume ratio that is more effective for heat loss in equatorial regions. Conversely the face was becoming proportionally smaller relative to the domed (850cc) cranium, however the primitive bony brow-ridge was retained and the lower jaw was still quite robust although the cheek teeth were smaller. These were the people who first dispersed within and without Africa, the first of the great human adventurers.

Adventurers, displaced persons, economic migrants or what?

Dispersal is a fundamental necessity for the survival of any species from flu viruses to salmon and humans. Some organisms and their gametes may never disperse over distances any greater than a metre or two, but for most species the production of progeny that never leave 'home' is a problem because of finite local resources.

There are of course many different strategies for dispersal; even within the primates the movement of the young beyond their natal (native) group varies depending upon the social structure of the group, the environment and territory they occupy and their mode of living. While the chimpanzees might provide a model for our most remote human relatives they are not so relevant to the situation in

which the early members of our genus found themselves in Central Africa at the beginning of Pleistocene times around two million years ago.

The first arms race

Biologically and socially, the African *Homo erectus/ergaster* people had moved a long way from their more ape-like ancestors and the environment was changing around them. The beginnings of a global climate change can be traced a long way back to early Miocene times.

Events such as the uplift of the Himalayas, which affected the production of the Southeast Asian monsoon, impacted upon Central Africa, which began to experience increasing aridity. By late Miocene times, around seven million years ago, global cooling and aridification resulted in the demise of some tropical forests and the evolution of grasslands along with grazing mammals in the Americas, Asia and Africa. With the beginning of Pleistocene times, around two million years ago, cooling produced the growth of ice sheets and glaciers beyond the polar regions. In Central Africa the growth of savannah grassland at the expense of forest produced a more varied landscape occupied by new grazing species such as buffalo, cattle, antelope, zebra-type horse relatives and rhinoceros.

There was an arms race between these mobile plant-eaters and their predators such as the cats and hyenas. For the grazers, safety in numbers plus an ability to keep moving over long distances was a successful adaptive strategy. However, its downside is

In early Miocene times, around 17 million years ago, Africa was almost entirely forested and extensive savanna grasslands were beginning to develop and spread as climates became drier. Early primates like *Proconsul* lived in the trees which provided them with food and safety from early carnivores such as the felid *Machairodus* which also preyed on other mammals such as the pigs and rhinos of the time.

Vegetation distribution for 18,000 BP climate

1   TROPICAL RAINFORESTS

2   DROUGHT-DECIDUOUS FORESTS AND DRY SEASONAL FORESTS

3   DROUGHT-DECIDUOUS WOODLANDS

4   TEMPERATE EVERGREEN SEASONAL BROAD-LEAVED FORESTS

5   SAVANNA

6   COLD-DECIDUOUS NEEDLE-LEAVED FORESTS AND WOODLANDS

7   COLD-DECIDUOUS BROAD-LEAVED FORESTS AND WOODLANDS

8   MEDITERRANEAN FORESTS AND WOODLANDS

9   EVERGREEN NEEDLE-LEAVED FORESTS AND WOODLANDS

10   MESIC GRASSLANDS

11   ARID GRASSLANDS AND SHRUBLANDS

12   DESERT

13   TUNDRA

14   POLAR DESERT AND ICE

Reconstruction of past climates and how they have changed depends upon a variety of data. These include direct evidence of glaciation and related processes preserved in deposits and landforms as well as fossils of animals and plants that have known climate tolerances. Combined with other proxy temperature information and mathematical modelling of past climates, it is becoming possible to map climates and vegetation for specific time intervals in the past.

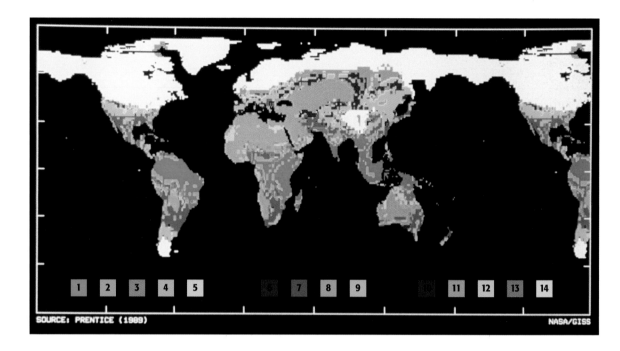

SOURCE: PRENTICE (1989)   NASA/GISS

that stomach size and weight has to be kept down to allow for speed of running, and reproduction becomes highly risky unless the offspring can get up on their feet and start moving immediately after birth – literally, within a few minutes.

Large group numbers also meant that the grasses that they depended upon could easily become overgrazed if the herd stayed in the same place for any length of time. In any event, many grasses are seasonal in their growth and consequently migration became a necessity of life. Obviously, as the herds of 'meat on the hoof' moved, so their predators moved with them.

By this time too, early human species such as African *Homo erectus/ergaster* had probably become active hunters and predators as well as scavengers, and they were perhaps the most adaptable of all. However, with marked differences in body size and strength between the sexes, they would have operated in hierarchical, male-dominated groups of no great size, probably 20 or so at most.

The final cause of what prompted African *Homo erectus/ergaster* to move is still not clear. Perhaps climate change combined with population increase put pressure on dwindling resources. Since they had no way of knowing what they were moving towards it is highly unlikely to have been simple curiosity or a kind of boy scout 'spirit of adventure'. More likely it was a case of 'when the going gets tough the tough get going'.

But they also seem to have possessed a very important innovation – the ability to make fire.

Fire power brings advantages

The possession of fire-making skills confers several notable advantages. It gives heat during cold tropical nights, affords protection against many predators and allows the cooking of meat. Cooking in turn helps to make protein more readily digestible and therefore helps release its energy sooner. It also kills potentially harmful bacteria and parasites in the meat.

Reliable evidence for the use of fire technology is hard to come by but there is some evidence for the early use of fire technology around 1.6 million years ago in the sediments of Swartkrans Cave, South Africa and Lake Turkana, north-eastern Africa associated with *Homoerectus/ergaster* remains. And for some time there has been considerable debate over evidence from *Homo erectus/ergaster* related hearths at Zhoukoudian in China, dated to around 500,000 years ago.

Recently, more secure evidence for the use of fire by *Homo erectus/ergaster* has been found in Israel and dated to around 790,000 years ago. Altogether the data does seem to support the idea that use of fire was an essential 'tool' for the first human-related migration out of Africa.

The archaeological evidence shows that by 1.8 million years ago African *Homo erectus/ergaster*-like people had for the first time moved out of the tropics and over 40° north, as far as Georgia. Other anatomically similar relatives were dispersing eastwards through India and on to Southeast Asia.

Armed with only simple stone tool technology comprising hammers, choppers, blades and basic hand-axes, plus the ability to make fire, they dispersed across an amazing variety of landscapes. It is possible that they kept as much as possible to coastal routes but we really have no information about this except that when they got to the Indonesian archipelago they must have acquired *some* seafaring skills. Even with lowered sea levels there are deep-water channels between some of the islands that could be crossed only by means of some kind of watercraft.

The archaeological evidence, especially of stone tools, shows a remarkable degree of conservativism in technology for over a million years until the beginning of late Pleistocene times around 100,000 years ago. The success of *Homo erectus* in Asia was remarkable and extensive both geographically and temporally, surviving until as recently as 100,000 years ago or less in Indonesia. Furthermore, the discovery of a new fossil human relative *Homo floriensis* indicates that this dwarfed descendant species of *Homo erectus* survived until around 18,000 years ago. This was long after the Neanderthals had died out in Europe (around 28,000 years ago) and long after the first modern humans had dispersed through Southeast Asia and Indonesia into Australia (by 50,000 years ago). Dwarfism of mammal populations stranded on islands is a well-known phenomenon especially in late Pleistocene times when the ice-age megafauna such as the mammoth was becoming extinct all over the globe.

Eurasian complications

While *Homo erectus* was surviving if not necessarily thriving in Asia, the African population of *Homo erectus/ergaster* was slowly continuing to change with increasing brain size. From around 900,000 years ago a new species variably called *Homo heidelbergensis/antecessor* can be recognised in both Central Africa, the European mainland and England.

The exact inter-relationships are far from clear but it is likely that this species gave rise to the Neanderthals in Eurasia from around 400,000 years ago. The Neanderthals had considerable success as a species in western Eurasia for well over 350,000 years, and showed some interesting advances in terms of behaviour and culture. They used fire for cooking meat, hardening the tips of wooden spears and preparing birch resin as a glue (see p. 63).

Some of them also buried their dead and made personal ornaments – attributes normally associated with *Homo sapiens*. The Neanderthals were also still in place when the first modern humans turned up in their home territories in western Eurasia around 32,000 years ago. At one time it was thought that modern Europeans may have evolved from the Neanderthals but we now know from both anatomical and genetic studies that this did not happen. The origin of modern humans lies back in Africa.

We are all Africans – the archaeological evidence

Around 400,000 years ago in Africa the *Homo heidelbergensis* people began to show changes, particularly in skull form, which began the trend towards *Homo sapiens*. The transition seems to have taken over 250,000 years and is not very well represented yet by skeletal remains in Africa, but there are some.

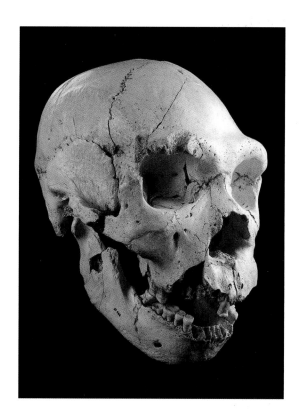

Armed with simple stone tools, and able to make fire, they were able to disperse across an amazing variety of landscapes.

This beautifully preserved skull from the Sima de Los Huesos at Atapuerca, near Burgos in Spain belonged to an elderly individual with worn and badly diseased teeth. Some 400,000 years old, the individual belonged to *Homo heidelbergensis,* a species that may have been ancestral to both the Neanderthals and ourselves, *Homo sapiens*.

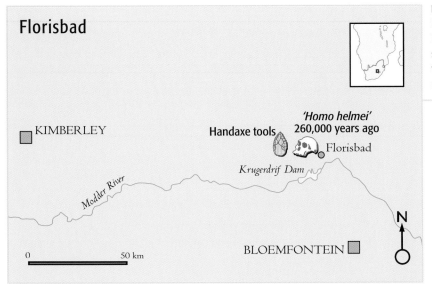

A skull found in 1932 at Florisbad in South Africa associated with late hand-axe stone tool technology was originally given the name *Homo helmei* and for a long time it was thought to be relatively young and perhaps contemporaneous with the Neanderthals in Eurasia. It does indeed show Neanderthal-like features in retaining a primitive browridge, receding forehead and large face. However, direct dating of one of the teeth in 1996 produced the surprisingly ancient date of 260,000 years, suggesting that it could well represent a link between *Homo heidelbergensis* and early *Homo sapiens*.

From the end of mid Pleistocene times, around 195,000 years ago, there are a few scattered African sites with important skeletal remains and evidence of advancing tool technology. The oldest remains come from sites such as Omo Kibish, in Ethiopia, and were found by Richard Leakey's team back in 1967. From here a number of fragmentary skeletal remains of three individuals were found, including those of a tall well built male whose skull has a distinctly modern looking chin.

At the time the remains were not thought to be very old, perhaps around 130,000 years according to relative dating by mollusc shells found with the bones. However, recent re-dating of some volcanic ash from sediments in which the fossils were found has provided a new date of 195,000 years ago, making these the oldest skeletal remains related to *Homo sapiens*.

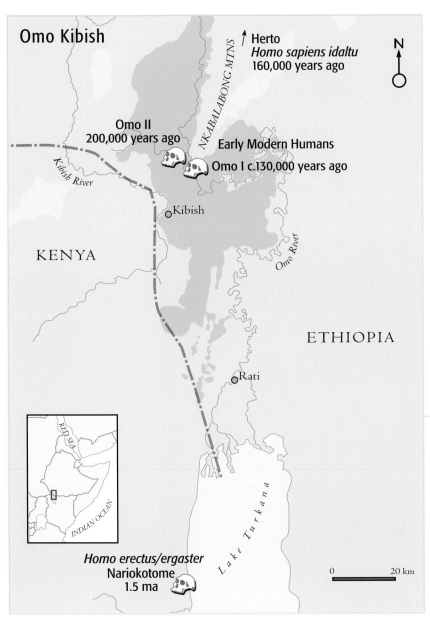

Omo Kibish – in 1967, Richard Leakey's team found the fossil remains of three individuals at Omo Kibish in Ethiopia. Omo I, dated to around 130,000 years old, has some distinctly modern traits such as a clear chin and divided brow. Omo II is slightly more primitive and is older, being dated at around 200,000 years old. Another well preserved skeleton from Herto, to the north, has been dated at around 160,000 years old and is the oldest well defined example of early modern humans so far discovered. It has been named as a new subspecies *Homo sapiens idaltu* but not all experts agree with the distinction. To the southwest, beside Lake Turkana, one of the best preserved fossil human skeletons was found in 1984. It belongs to a tall eight year old boy, identified as the African species *Homo ergaster* (synonymous with Asian *Homo erectus*).

Until this recent revision of the Omo Kibish dates, the oldest *Homo sapiens* skeletal material was that from Herto, also in Ethiopia, and dated at around 160,000 years ago, which is perhaps the oldest of the early *Homo sapiens*-related remains yet found.

The Herto remains include the fossilised skulls of one young individual and two adults associated with stone tools with features normally identified as belonging to both Acheulian and Middle Stone Age technologies. In addition, the butchered remains of hippo carcasses were found.

The best-preserved adult skull has a brain capacity of about 1450cc and all the cranial bones have cut marks indicative of mortuary practices. The latter include evidence of defleshing, the removal of the jaw, blood vessels, nerves and muscles. Tim White and the other scientists who first described the Herto remains in 2003 think that they show enough slight differences from modern *Homo sapiens* to justify placing them in a new subspecies *Homo sapiens idaltu*. However, as usual other experts disagree, especially in view of the re-dating of the Omo Kibish remains to which the Herto finds show considerable resemblance.

Several early *Homo sapiens*-type skulls have also been found in North Africa. These include those of two adults from Jebel Irhoud in Morocco and dated at around 150,000 years old. The presence of Middle Stone Age tools in central and West Africa indicates that early modern humans were also present there, but unfortunately we do not have any skeletal material to know what they looked like. Overall, there is now good archaeological evidence for an African origin of modern humans from at least 195,000 years ago and with an ancestral connection to *Homo heidelbergensis* stretching back over 400,000 years.

The Levantine corridor

The route northwards out of Africa is not an easy one. It crosses the Tropic of Cancer with its extreme climate that results in the heat and aridity of the Saharan and Nubian deserts bounded by the Red Sea to the east. There is of course the Nile Valley and that may well have provided a line of exit but the archaeological evidence so far points to a more coastal route along the shore of the Red Sea where there is a good supply of readily accessible seafood.

There are middens (refuse sites) at Abdu in coastal Eritrea, which contain the empty shells of edible molluscs and the butchered bones of small game. Middle Palaeolithic stone tools dated at around 125,000 years old have also been found, but no human-related skeletal remains. North from here there is just the narrow isthmus that connects present day Egypt to the arid Sinai Peninsula and then north along the Levantine coast (today's Israel and Palestine).

To the east lie the inhospitable deserts of Syria and Saudi Arabia. The narrow coastal plain of the Levantine is bounded by the Mediterranean to the west and a line of hills and mountains to the east. In places there are rocky prominences made of limestone and riddled with caves that provided ideal shelters and lookout sites for early modern humans and their distant relatives, the Neanderthals.

The Levantine Corridor – Evidence for the early movement of modern humans (H) out of Africa has been discovered in the so-called 'Levantine Corridor' sites of Qafzeh, Tabun and Skhul. When originally found, they were thought to be around 40,000 years old but redating has pushed the dates back to between 120,000 and 80,000 years ago. Younger remains found at Amud and Kebara belong to Neanderthals (N) and are dated to between 60,000 and 50,000 years old. The implication is that there seems to have been some crossover here between Neanderthal and early modern human populations.

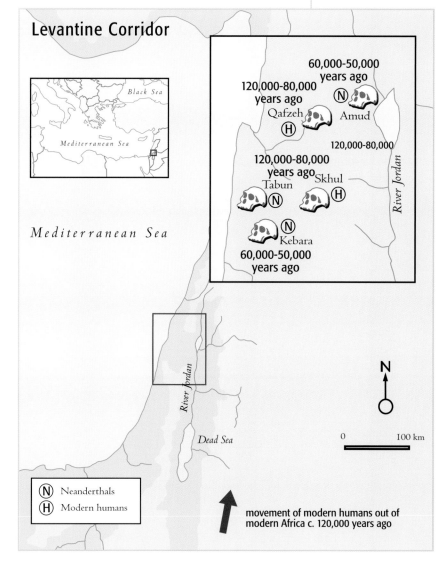

In the 1930s a series of cave excavations in the semi-arid Wadi el-Mughara near Mt Carmel revealed some well-preserved human-like remains and stone tools. The Anglo-American excavation team uncovered the remains of some 10 people at Skhul; men, women and children who had all apparently been purposefully buried beneath a metre or more of cave floor sediment. Nearby at Tabun, human-related bones were less common and consisted of the partial skeleton of a woman. The bones of gazelle and fallow deer were all found at a somewhat greater depth below the cave floor. The accompanying stone tools in both cases were Middle Palaeolithic (Levallois-Mousterian) in style and so the initial conclusion was that all the human-related remains were probably those of contemporary groups derived from the same regional population.

Excavation at Kebara cave, one of a cluster in the so-called 'Levantine corridor', has uncovered one of the best preserved Neanderthal skeletons and the only one to preserve a nearly complete pelvis. The bones and ribs are robust and belonged to a powerfully built individual, only the skull and lower limb bones are missing.

Study of the detailed anatomy showed however that there was considerable variation. The Skhul people looked decidedly modern while the Tabun woman had a more Neanderthal-type skull, although her frame was quite lightly-built for a Neanderthal. And so it appeared that they might be hybrids between the Neanderthals and early modern humans, but the main problem was that the only available dating depended upon identification of the tools as Middle Palaeolithic in age.

Then excavation of another cave at Qafzeh near Nazareth revealed firstly the remains of another seven individuals; a further 14 were recovered after WW2. Again they looked quite modern and had been buried but were associated with Mousterian tools that are normally linked to the Neanderthals. More Neanderthal remains were found at other cave sites: at Amud a skeleton was found right at the top of a sequence of Middle Palaeolithic sediments and tools, while one of the best Neanderthal skeletons known was found at Kebara — only the skull was missing.

A complex story

Eventually, comparison of all the bones showed that the remains at Tabun, Kebara and Amud were all typically Neanderthal while those of Skhul and Qafzeh were more modern looking. Inevitably the conclusion was that the latter were more recent and the former had to be more ancient, but there were hints from the accompanying animal fossils that the story was more complex. However, at the time it could not be checked by radiocarbon dating (see box p. 182) because the material was too old.

The traditional radiocarbon dating methods were restricted to material less than 40,000 years old and even the advances of accelerator mass spectroscopy (AMS) did not extend the method back more than 70,000 years. However, by the mid 1980s thermoluminescence (TL) dating and electron spin resonance (ESR) dating (see p. 182) became possible and when they were applied to burnt flints and animal teeth from Kebara they produced a date of around 60,000 years.

The big shock was the TL date of the early modern human burials at Qafzeh because it was so old – dating back to 92,000 years ago. Thus the remains were three times older than those of the earliest modern human remains in Europe and 32,000 years older than the Kebara Neanderthals. Then, to cap it all, the Skhul modern human remains were indirectly dated (by testing associated animal teeth found at the site) at between 100,000 and 80,000 years old by the ESR method.

Finally, the Tabun Neanderthal woman was re-dated to between 200,000 and 80,000 years old and so may have been contemporary with the Qafzeh and Skhul modern humans – in other words the previously accepted chronology was effectively turned upside-down.

Evidently there was a continuous flux of populations of both Neanderthals and modern humans occupying the region over a long period, with the modern humans coming up from Africa and reaching here by at least 90,000 years ago. The 'to-ing and fro-ing' was probably related to changing climate. From around 200,000 to 120,000 years ago there was a cold glacial period followed by a briefer warm phase until around 75,000 years ago when there was another glacial period.

Moving into Europe and Asia

There is plenty of archaeological evidence from both bones and stones for the presence of modern humans in Europe. Despite their initial discovery in the early part of the 19th century, finds such as the Paviland burial in Wales (see p. 73) could not be recognised for what they were because of the prevailing mindset which accepted a literal interpretation of the Judeo-Christian bible, and in particular the Old Testament account of the special creation of humans.

Over the centuries many attempts were made to develop a dated history for this and events such as the Noachian Flood from biblical and other emerging sources of early history. The generally accepted date for Creation was around 4004 BC but many geologists were already questioning the older chronology and ideas about the age of the earth.

Then in 1868 bones representing the partial skeletons of three individuals were found buried in sediment below a cliff overhang, which formed a natural rock shelter known as Cro-Magnon near Les Eyzies in the Dordogne region of France. The remains were accompanied by Upper Palaeolithic stone tools, some pierced shells and the remains of

animals such as mammoth and reindeer. The site was excavated by the French archaeologist Louis Lartet (1840-1899) who realised that, at last, this was incontrovertible evidence for the co-existence of humans with extinct 'antediluvian' animals. We now know that the site records the presence of modern humans in Europe around 30,000 years ago.

Since then innumerable other sites have been found and dated wherever possible, with some such as Vogelherd in southern Germany apparently dating back as far as 32,000 years ago. However, it has recently been realised that there are significant calibration problems with radiocarbon dates, which can fluctuate in a non-linear fashion.

Re-dating of many European skeletons has tended to make them younger than previously thought and certainly none seem to be any older than 32,000 years. The skeletons of these earliest modern humans in Europe, often referred to as the Cro-Magnon people, show that they were distinctly different from the incumbent Neanderthals. The incomers had typically North African body shapes with long limbs and small faces below high foreheads. By contrast the Neanderthals were much more heavily built with shorter muscular limbs and big faces with low foreheads and bony brow ridges.

However, what the Cro-Magnons lacked in terms of strength they compensated for in their new technology. They were developing a greater variety of stone tools, using bone and ivory to develop new kinds of items such as barbed harpoons, spear throwers, needles and personal ornaments.

They buried their dead, produced art and were developing counting and perhaps calendar systems. But there are much older remains far to the east of Europe, which show that these African modern human emigrants found it easier to travel in subtropical climes.

Some of the most remarkable evidence for the migratory, or more correctly we should say dispersive tendencies, of modern humans comes from Australia. The Willandra Lakes region of New South Wales has a number of early settlement sites most important of which is known as Mungo 3.

A virtually complete skeleton, dated at around 40,000 years old, was found buried there with a dusting of red ochre, presumably as a ritual gesture. Originally, this semi-arid region was a fertile

The Cro-Magnons were deficient in terms of strength but compensated for it by their new technology. They were developing a greater variety of stone implements, using bone and ivory to develop new kinds of tools.

Willandra Lakes

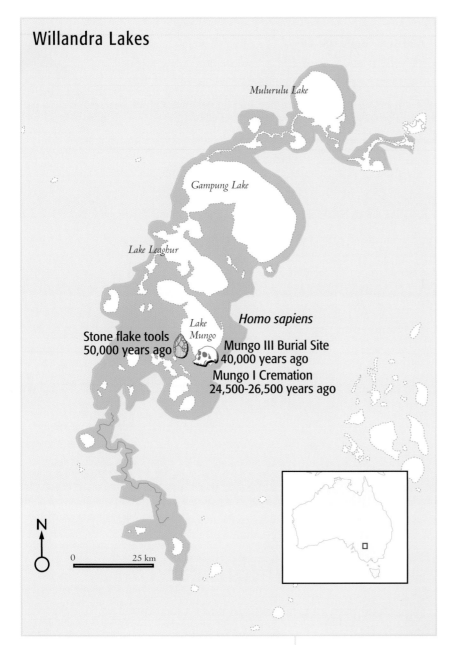

Mulurulu Lake

Gampung Lake

Lake Leaghur

Lake Mungo

**Stone flake tools
50,000 years ago**

Homo sapiens

**Mungo III Burial Site
40,000 years ago
Mungo I Cremation
24,500-26,500 years ago**

N

0   25 km

Mungo's Firsts

Lake Mungo in the Australian state of New South Wales contains Australia's oldest human remains, Mungo III. The remains were found in the world's oldest ritual ochre burial, from which the oldest human mitochondrial DNA has also been obtained, plus the first recorded cremation (Mungo 1).

But ever since the first find was made in 1969 (Mungo 1) there have been arguments over dating. To begin with radiocarbon dates placed Mungo 1 at around 20-26,000 years ago whilst Mungo 3 has been variously dated at 30,000, 42,000-45,000 and around 62,000 years. Since these latter dates are beyond the normal range of radiocarbon dating, other methods have been employed to try and obtain more reliable dates.

Recently (in 2003), four different dating laboratories independently dated the same samples of sediments from the immediate surroundings of the finds, using optically stimulated luminescence (OSL) and came up with similar results for the first time.

From 25 optical ages it appears that both Mungo 1 and 3 burials were made around 40,000 (+/- 2,000) years ago. However, there is also evidence that humans entered the area between 50-46,000 years ago, at about the same time that they arrived in northern and western Australia and coincident with the first extinctions of the Australian megafauna. The sediments also record fluctuating and deteriorating climates between 50-40,000 years ago ranging from periods when the lakes were full to much drier conditions.

Willandra Lakes – In 1969 female human remains were found buried at Lake Mungo, one of 17 lakes within the Willandra Lakes World Heritage Area, 987 km west of Sydney, Australia. Dated at between 24,500 and 26,500 years old they are the oldest known cremation. Five years later in 1974, a male skeleton of a modern human was found nearby covered in red ochre and purposefully buried in a shallow grave, and this has been dated at around 40,000 years old. Stone tools found nearby are even more ancient at around 50,000 years old. They show that modern humans originating from Africa

had reached Australia by this time. The original lake was full of fish and about 10m deep and the surrounding area was occupied by many of Australia's extinct marsupial animals such as giant kangaroos and 'Tasmanian tigers'.

'Mungo man' was probably buried approximately 40,000 years ago, when humans had been living in what is now the Willandra Lakes area for some 10,000 years. The discovery and dating of this skeleton has helped to prove the 'Out of Africa' theory.

vegetated area of lakes and swamps occupied by a variety of game such as kangaroos, plus fish and clams, which provided a plentiful food supply. For these people to have arrived in this south-eastern part of Australia at least 40,000 years ago they would have had to have entered the continent considerably earlier from the north west after island hopping through the Indonesian archipelago. That means that even with lowered sea levels they would have needed some kind of raft to cross the few deepwater straits between certain islands. Unfortunately we have no archaeological record of such craft.

The remarkable miniature beings of Flores

It is also intriguing to think that on their way through the Indonesian islands they may well have encountered another human-related species *Homo floresiensis*. This recently discovered dwarf species, standing no more than a metre (3 ft) tall, is thought to have been closely related to the *Homo erectus* people who were the first to disperse beyond Africa nearly two million years ago. *Homo floresiensis* managed to reach the Indonesian islands by around 1.8 million years ago and seem to have survived until perhaps as recently as 55,000 years ago. There may even have still been a few stragglers around when modern humans first passed through the islands, but this we do not know.

However, since the discovery of *Homo floresiensis* in 2004 by a joint Indonesian-Australian team we *do* know that these remarkable little humans were alive at the time when modern humans first entered the region. They must have had a large and sufficiently viable population to survive at least until 18,000 years ago – which is the most recent evidence we have for them.

A few experts have claimed that the find is nothing more than that of a pygmy or mutant modern human child who suffered from microcephaly – a genetic condition in which the head and brain do not develop. However, tomographic scanning of the 400cc, chimpanzee-sized brain surface topography shows that it has some distinct features normally associated with the brain of *Homo erectus* and there are other features of the skull which are more like those found in the *erectus* people than in modern humans. Finally, they are not the only species to have been dwarfed on

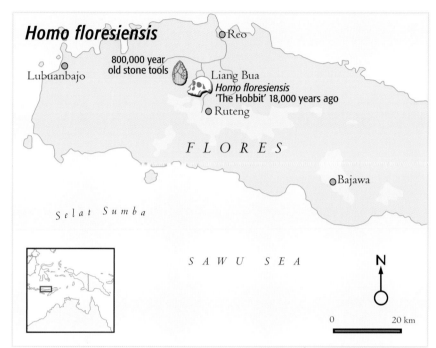

Flores, for they shared the island with extinct dwarf elephants (*Stegodon* spp.) as well as some predatory komodo dragons.

It has long been known that communities of animals stranded on islands with finite food resources can respond to such constraints over time by becoming dwarfed and thereby surviving on less food. For instance, on the islands of Sicily and Malta, elephants and hippos barely 1m (3ft) in height became the dwarfed descendants of 4m (13ft) ancestors within some 5,000 years of the end of the last ice age.

But the existence and survival of these little people does raise some important questions about the so-called 'cerebral rubicon' for membership of our genus *Homo*. For many years the baseline measure for humanness was a 700cc brain. Louis Leakey lowered the threshold to 600cc in order to include his species *Homo habilis*, however not all experts have agreed and some continue to place his species in the smaller brained genus *Australopithecus*. Nevertheless, with a mere 400cc brain, little *Homo floresiensis* is way below either measure and even when allowances are made for its small body size, it still falls well below the accepted definition of our genus.

The problem is that we do not know what effect this small brain had on the behaviour and lifestyle of these extinct Floresians. Did it have a significant

Homo floresiensis – Found in 2003 on the Indonesian island of Flores, the 18,000 year old skeleton of a metre-high extinct human relative, named as *Homo floresiensis*, has upset received opinion about what constitutes a human species. Nicknamed the 'hobbit', the total remains belong to more than one individual and are interpreted as those of a dwarf species, perhaps descended from *Homo erectus* people who occupied the region from around 1.5 million years ago. Dwarfism is not uncommon especially where populations of animals are 'trapped' within the relatively confined space and resources of islands. Stone tools, dating from 800,000 years ago until some 12,000 years ago, show a consistency of style and suggest that they may have been made by *Homo floresiensis* people.

With a mere 400cc brain, little *H. floresiensis* falls well below the accepted definition of our genus.

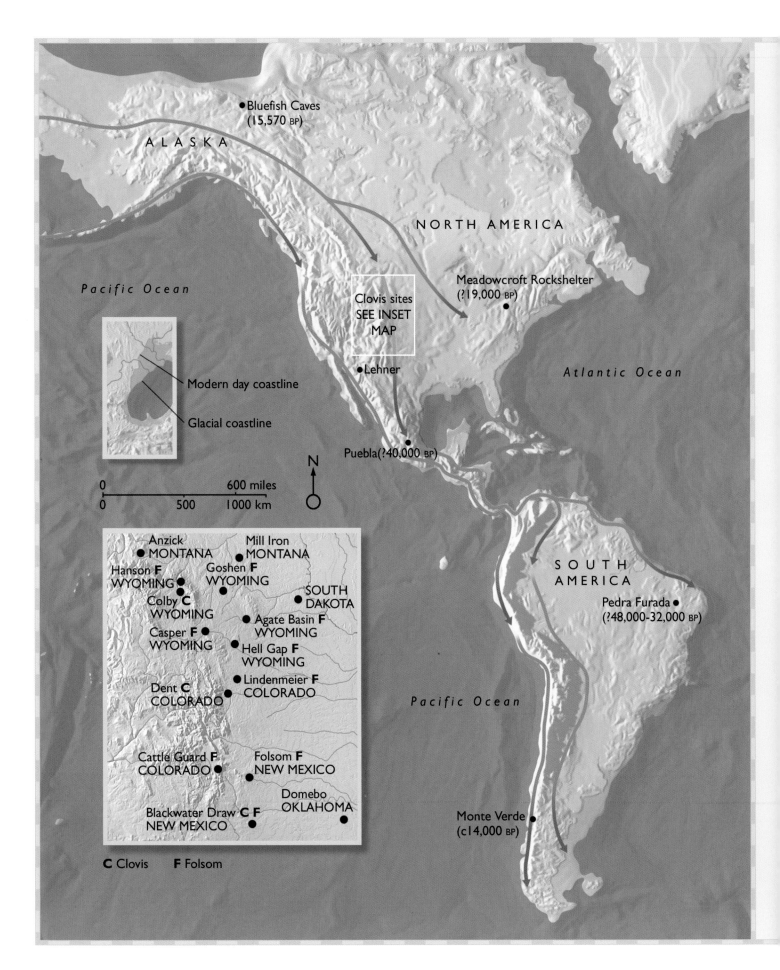

Bluefish Caves
(15,570 BP)

ALASKA

Pacific Ocean

NORTH AMERICA

Meadowcroft Rockshelter
(?19,000 BP)

Clovis sites
SEE INSET
MAP

Atlantic Ocean

Modern day coastline

Glacial coastline

●Lehner

Puebla(?40,000 BP)

N

0		600 miles
0	500	1000 km

Anzick
MONTANA

Mill Iron
MONTANA

Hanson **F**
WYOMING

Goshen **F**
WYOMING

Colby **C**
WYOMING

SOUTH
DAKOTA

Agate Basin **F**
WYOMING

Casper **F**
WYOMING

Hell Gap **F**
WYOMING

Lindenmeier **F**
COLORADO

Dent **C**
COLORADO

Cattle Guard **F**
COLORADO

Folsom **F**
NEW MEXICO

Domebo
OKLAHOMA

Blackwater Draw **C F**
NEW MEXICO

C Clovis   **F** Folsom

SOUTH
AMERICA

Pedra Furada ●
(?48,000-32,000 BP)

Pacific Ocean

Monte Verde
(c14,000 BP)

Settling in the Americas

The timing of settlement in the Americas by modern humans is one of the most contentious topics in archaeology. During the last Ice Age much of North America was covered with a vast ice sheet that extended south to the region of the Great Lakes that were formed when the ice melted. The traditional view is that modern human hunters first entered North America from Asia across the Bering landbridge when lowered sealevels exposed the Bering Strait as dry land during the last Ice Age. Genetically and linguistically, Native American peoples are related to Native peoples of eastern Asia which was settled by modern humans perhaps as long as 40,000 years ago.

Entry to North America was blocked by a vast ice sheet but there may have been an ice-free corridor just east of the Rockies or they may have taken a coastal route. Well documented artefacts in the form of finely modelled Upper Palaeolithic-like stone spear points have been found at a number of localities in the Mid-west. They are known as the Clovis and Folsom cultures and date to no more than 13,500 years ago.

However, in recent decades a number of other sites have been found scattered from northern Alaska to Chile in South America and are claimed to be pre-Clovis in age with some dating back to around 19,000 years ago (Meadowcroft) and even perhaps as far back as 35,000 years ago (Pedra Furada in Brazil) or 40,000 years ago at Puebla in Mexico but the dating is vehemently disputed. Nevertheless the evidence is definitely mounting for an earlier settlement of the continents of the Americas.

effect on their cognitive abilities and if not, why not? Recent analysis of the brain shape with the size and position of the frontal and temporal lobes indicates that despite its overall small size *Homo floresiensis* was capable of the kind of higher cognitive processing normally associated with our genus. It will be intriguing to see if we can get any answers from the archaeological record of stone tools associated with the bones.

Into the Americas

One of the most contentious issues in the story of how modern humans spread around the globe is the question of the Americas. No sooner had the first European visitors to the Americas come across the living native inhabitants than questions were raised about who they were and how they had arrived there. Inevitably, there were at first some fanciful answers to these questions. For instance in the 16th century some Spaniards thought that they must be one of the lost tribes of Israelites who had stumbled across some northern land connection into the continent.

For early 20th century scientists, the question of *how* the native Americans got there was not too much of a puzzle as there is only one really feasible route and that is from Siberia via the Bering Strait and into Alaska. Today the Bering Strait is an inhospitable seaway but scientists were – and are – well aware that during the ice ages there were many times when a lower sea level would have exposed a land connection between Siberia and Alaska. This land bridge is known as Beringia. It acted as a two-way freeway for both quadrupedal animals and bipedal humans, who probably discovered the route by following the game they hunted. Without this connection the fauna of the Americas would be much more ancient and strange, and dominated by marsupial mammals like those of Australia. The big question is: when did modern humans first move into the great northern and southern continents of the Americas?

There is a scattered archaeological record, which mostly consists of virtually indestructible stone tools. However, they are often difficult to date and need to be buried in stratified sequences that contain associated datable materials such as bones, to which radiocarbon methods might be applied, or eruptive volcanic rock material that can be dated by other radiometric methods.

Failing this, reliance has to be placed on the traditional methods of comparative stratigraphy using fossils (see box on p. 86). It was not until the 1930s that evidence was first found of stone tools in the form of well-fashioned distinctive points associated with the bones of extinct mammals, especially bison. Called the Clovis and Folsom cultures they were clearly mobile hunters of big game that included mammoth, bison and horse. This association pointed to an occupation of North America that was close to the end of the last ice age.

The problem with this is that the northern part of the continent was covered in a huge ice sheet that extended down across the modern border between Canada and the US to the region of the Great Lakes. And, along the western margin lay the mountains and glaciers of the Cordillera and Rockies. How could these hunters have made their way across such an ice sheet? It seemed completely impossible.

By the 1960s, dating of the oldest Clovis culture artefacts extended no further than around 13,500 years ago and more detailed mapping of the ice-age climate change and the associated growth and melting of the ice sheet revealed the possibility that there may have been an ice-free corridor between the western mountain glaciers and the continental ice sheet. However, the timing of glaciation precludes an ice-free corridor between 22,000 and 13,500 years ago – so how did the first Americans get into the continent?

New evidence emerges

In recent decades new evidence has emerged of a significantly earlier phase of immigration and of the possibility of an ice-free coastal route into the continent. Excavations in the Bluefish caves in the Yukon have revealed many small flaked stone artefacts called microliths, and bones of a variety of animals such as mammoth, bison, horse and caribou that possibly date back to 15,750 years ago. The bones show signs of having been chewed by animals such as wolves and the caves were probably used by a variety of animal predators and scavengers.

Some of the bones may have been fashioned by human hands and so it is possible that migrating

modern humans also used the caves. Further into the continent there are also a number of other sites such as Meadowcroft in Pennsylvania that have variously been claimed to predate the Clovis culture, but there are problems of dating with most of them.

However, there some sites further south which are seriously disturbing the accepted view and have been the subject of a great deal of controversy. First there is Monte Verde in South America on the Chilean coastal plain nestling between the Andes Mountains and the island-studded Pacific coast.

In the late 1970s, American archaeologist Tom Dillehay followed up a local discovery of bones and realised that here was a site of great interest. So began a 20-year programme of detailed excavation. The marshy riverbank site had preserved the remains of a small settlement of timber-framed huts that were originally covered with animal skins, fragments of which were still preserved in the muddy sediments. There were also the remains of hearths, cooking debris and other rubbish, wood tools and flaked stone artefacts, and animal bones including those of the extinct mastodon. These have been dated at between 14,500 and 13,800 years ago.

The site is nearer the Antarctic Circle (some 3,500km or 2,200 miles to the south) than the Arctic Circle (around 15,000km or 9,400 miles) to the north, but the only way that the occupants of Monte Verde could have got there is from Beringia, the same as the other early settlers. One question is this: how long might it have taken them to make their way south?

Although they were mobile hunters, to have sustained a viable population they would have been accompanied by women and children and are unlikely to have travelled much more than a few kilometres each year. Supposing for sake of argument they travelled at an average speed of 5km (3 miles) per year, it would have taken them 3,000 years to reach Monte Verde and their starting time would have been between 17,500 and 16,800 years ago in Beringia. However, Beringia was in the grip of the last glaciation at that time and it is unlikely that they could have moved through the region.

The only answer that seems to fit is that they reached the continent *before* the onset of that glaciation, and that would have had to be around

New evidence emerged of a significantly earlier phase of immigration and of the possibility of an ice-free coastal route into the continent.

22,000 years ago. Interestingly, the Monte Verde site also has some problematic evidence of an even older settlement that has been dated to around 33,000 years ago by radiocarbon dating of scattered pieces of charcoal. Dillehay and his colleagues are themselves sceptical about these dates, however.

But there are two other sites of human occupation in the Americas from which dates of around 40,000 years and more have been obtained – Pedra Furada in north-eastern Brazil and most recently, Puebla in Mexico.

The South American site became headline news in 1986 when the Brazilian archaeologist Niede Guidon published an article in the international science journal *Nature*. She claimed that several hundred stone tools found at Pedra Furada recorded the oldest known human site in the Americas.

They were dated from associated charcoal by radiocarbon methods at between 48-32,000 years ago. The site also records a succession of younger occupations including two hearths dated at 32,000 years old, a rock art tradition from at least 12,000 years ago along with human remains.

Since then new radiocarbon dates have been obtained using more sophisticated accelerator mass spectrometry (AMS) because there were questions over possible contamination. Repeated dating of the original charcoal samples has been undertaken by the Australian National University in Canberra. Five samples returned dates that are even older than the first ones at more than 56,000 years and beyond the limits of the method. Another two samples gave more finite dates of 53,000 and 55,000 years.

From these, the investigators are now claiming that humans have occupied the site from 60,000 years ago. Critics still claim that the charcoal could have been derived from natural fires rather than human generated ones. It remains to be seen whether different dating methods will verify these claims and whether they will become more widely accepted. The discovery of yet more sites dating back well beyond 20,000 years will undoubtedly lead to a reassessment of sites such as Pedra Furada. Puebla in Mexico is one such site.

The footprints of Puebla

The discovery of some human footprints in Mexico has really put the proverbial 'cat among the pigeons' of American prehistory. British scientists discovered the prints in September 2003 at the bottom of an abandoned quarry near Puebla in central Mexico.

Excavations have shown that altogether some 270 prints have been preserved, of which two-thirds are human and the rest belong to other animals including mammoths and other extinct mammal species of cattle, deer and camel. There is no doubt that the human-related prints were made by fully modern upright-walking modern humans. From their different sizes, the group included both adults and children as young as five or six years old. And, analysis of the stride length between steps and the size of the prints indicates that they had a height range of between 1.14 and 1.83m (3ft 9in and 6ft).

The footsteps were originally impressed upon volcanic ash deposited around a lakeshore from the eruption of a nearby volcano. Reactions between chemicals within the ash caused it to harden off and preserve the hollow imprints (in a similar way to the preservation of the Laetoli prints, see p. 149), which were subsequently infilled with other sediment as part of the normal lakeshore sedimentation. Subsequent preferential weathering of the relatively softer and younger sediment has resurrected the original prints.

The association of the human prints with those of extinct mammals shows that the prints are at least 10,000 years old because that is when the last of the mainland American mammoths died out. However, more accurate dating of the prints has had to await more sophisticated techniques such as optical stimulated luminescence (OSL, see p. 182) and radioisotope dating. Scientists from Oxford University have now done this – and have come up with the startling date of around 40,000 years ago.

If this is correct and can be corroborated by other dating laboratories, then the story of the colonisation of the Americas will certainly have to be revised. As we have seen, at present the generally accepted date for the first colonisation is around 13,500 years ago. Although, if the dating of the Chilean site of Monte Verde is accepted, then it has to be pushed back to at least 14,500 years ago. This is controversial enough but the idea of having to add another 25,500 more years and push colonisation back to 40,000 years ago is going to be very hard to sell.

More recently, American geochronologists using a different technique have dated the footprint layer as around 1.3 million years old. There is a big problem – are they really human footprints, is the American date wrong, or did *Homo erectus* spread a long way beyond Southeast Asia?

CHAPTER TWO

Preserving our prehistory – what the archaeological record tells us

The vast majority of archaeological information about our most ancient human relatives is confined to a rather fragmentary record of bones, teeth and associated artefacts such as stone tools, spear heads and the like. The scientific reasons for this are quite complex but are now clearly understood – they are to do with chemical and geological processes which impact upon the burial and preservation of any vestiges of life in terrestrial environments. The fact is, in a natural environment full of scavengers and microbes, the 'economy of nature' tends to recycle any materials that are potential food for some life form or other.

An archaeologist excavates a human fossil at the Gran Dolina site in Sierra de Atapuerca in Spain. The objects and fossils at this site date back 400,000 years.

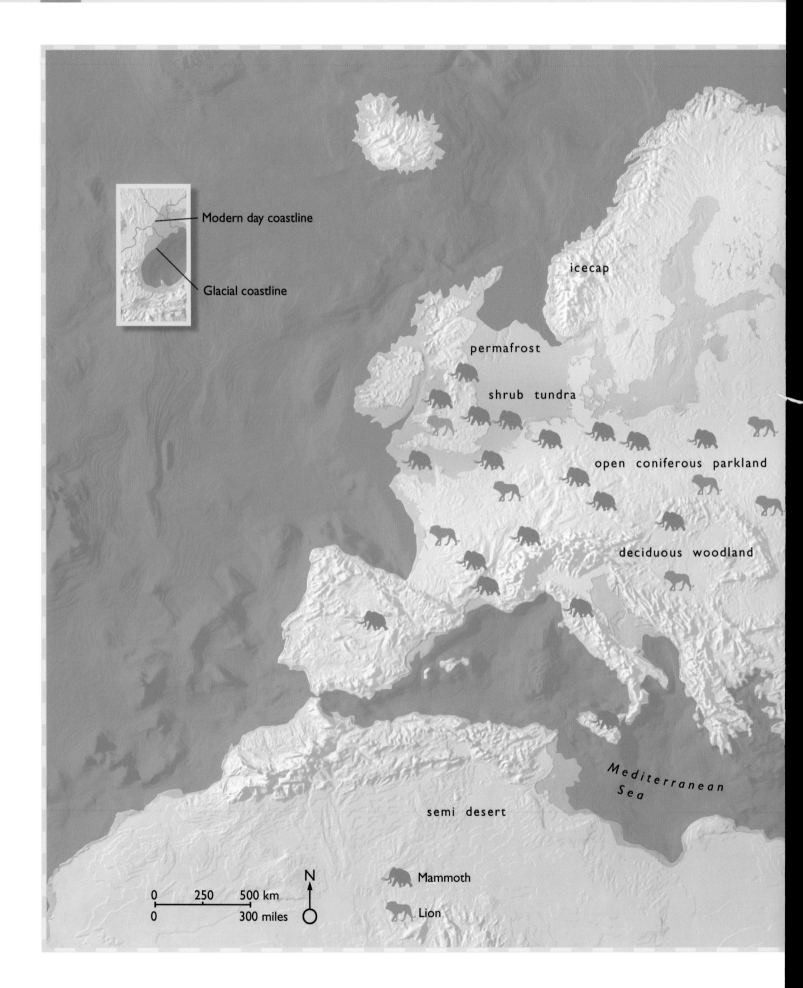

Modern day coastline

Glacial coastline

icecap

permafrost

shrub tundra

open coniferous parkland

deciduous woodland

Mediterranean Sea

semi desert

N

0   250   500 km
0   300 miles

Mammoth

Lion

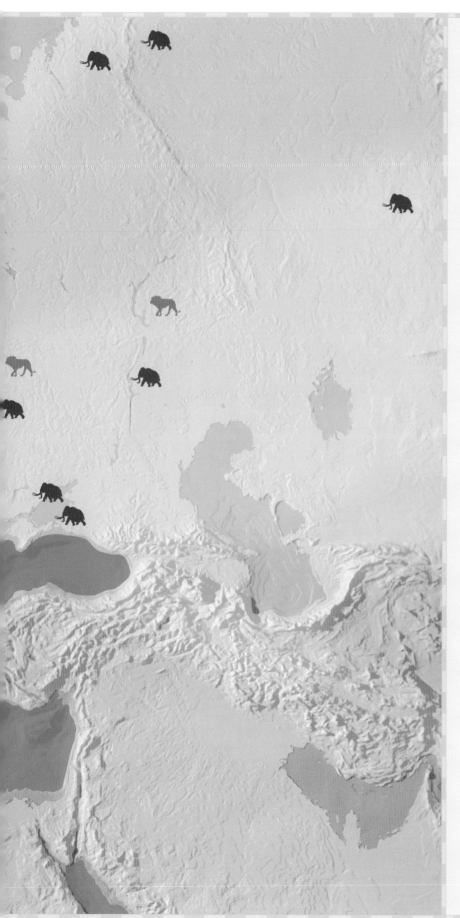

The Ice Age megafauna

Today it is hard to believe, but until around 12,000 years ago Europe and Asia was populated by a diversity of large mammals (ranging from elephants and rhinos to horses and hippos). Collectively, they are known as the Ice Age megafauna and are now virtually extinct, only the reindeer, wild horse along with a few European bison, bears and Siberian tigers remain, although there are plans to try and recreate an Ice Age Park in Siberia, it can never be the same as the real thing was. Our ancient human relatives were confronted by the Eurasian equivalent of an African Game Park. There is a long running controversy over what exactly caused the extinction of the Ice Age megafauna: climate change or human hunting or some combination of the two. At the moment our ancient relatives are held to be mostly responsible with a little 'help' from climate change.

The most iconic of the extinct animals of the Ice Ages is the woolly mammoth. Despite its name, mammoths were no larger than elephants today but they did have spectacularly large and curved tusks and were well adapted to cold and dry conditions in open woodlands and grasslands. They were very successful animals with herds roaming over large parts of Eurasia. Once mature, they had few natural enemies, except perhaps for lions, until human hunters appeared. Although most mammoths were extinct by around 11,000 years ago some survived on the offshore Wrangel Island in the Arctic Ocean until around 4,000 years ago.

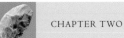

Extraction and amplification of fragmentary mammoth DNA shows that they are most closely related to the Asian elephant and diverged around 6 million years ago.

Naturally, this means that for there to be any hope of long-term preservation in the archaeological record, the remains have to be trapped either by some natural process such as rapid burial within sediment, or purposefully buried by human hand. As we shall see, purposeful burial is a conscious act associated with the advent of a sense of identity and 'the other' and formalized by religion. Although there are some Neanderthal burials, the practice is mostly associated with our species *Homo sapiens* and the last 80,000 years of our history.

Most human-related artefacts dating back to prehistoric times consist of stone tools and marks made upon rock – because these inorganic materials tend to be preserved. By comparison, softer organic materials such as animal skin clothing, wooden objects and even ivory tend to degrade over periods of more than a few thousand years and hence have been preserved only rarely.

However, there are certain environments with exceptional qualities of preservation in which more fragile organic materials may be found. One of the most fragile of these is body tissue and, most important of these are biomolecules of DNA. The hope of recovering very ancient DNA, promoted by science fiction epics such as *Jurassic Park* has, unfortunately, not been upheld by science fact. There were a number of reports, even in the scientific press, of the recovery of DNA tens of millions of years old from insects embedded in amber. But they were all the result of somewhat shoddy science where no effort was made to independently duplicate the results – and as it turned out, all were spurious results from contamination of the samples.

The problem is that although the sophisticated technology of amplification of fragments of DNA is now so sensitive that it can pick up the smallest trace, which is why the process has become such a powerful forensic tool (perhaps most notoriously with the identification of the owner of some dried-out DNA on a certain blue dress), it cannot perform miracles.

Ancient DNA has been recovered since the 1960s from archaeological material, most often part of museum collections. The sources have usually been dried tissues no more than a few thousand years old, such as the skin of the extinct horse-like quagga and

It it is generally only the most indigestible and resistant body parts that are preserved for thousands of years – teeth and dense bone, and of course, stone artefacts. But all of these, even the latter, while they are lying on the surface are exposed to the elements and are susceptible to the effects of natural destructive processes as a result of wind, rain and rough abrasive handling by rivers. It is only when these artefacts become buried beneath the surface of the ground that they are protected from these forces.

an Egyptian mummy. But since then the science of DNA recovery has developed enormously, along with an understanding of the environmental constraints on the preservation of ancient DNA. The ideal conditions are cool, dark and dry where there is no microbial activity, and instances of these conditions coming together are rare. Consequently, even bones lying out on the ground surface or even in shallow graves do not retain any trace of recoverable DNA for very long. In addition, most equatorial and even temperate climate zones are destructive to DNA.

Unfortunately this means that the chance of recovering any DNA from our African ancestors is extremely remote. However, the recovery of 40,000-year-old mitochondrial DNA from the Mungo 3 burial in Australia shows that it is possible in certain very dry and hot conditions. Even relatively young bones such as those of Indonesia's little *Homo floresiensis* people which are only some 18,000 years old, do not seem to have retained any recoverable DNA, despite being buried in a cave. Only in cool high latitudes or at very high altitudes in the tropics such as the Andes are conditions conducive to the preservation of DNA. The mummified child victims of Peruvian Inca sacrifices are exceptionally well preserved high up on the dry 'rain-shadow', lee-side

of the mountains where they were quickly and effectively freeze-dried by sub-zero wind-chilled air.

The oldest of ancient fragmentary DNA to be recovered so far is around 80,000 years old and comes from mammoths preserved in the frozen ground of the Siberian permafrost up in the Arctic Circle. Unfortunately, however, the possibility of finding modern human remains similarly frozen and

Unfortunately, the chances of recovering any DNA from our African ancestors is extremely remote.

Woolly mammoths preserved in the Siberian permafrost were first found by Siberian tribespeople centuries ago. This rather pig-like baby mammoth, whose trunk has been lost, was found in 1988. The rather fanciful 19th century engraving above assumes that they were just like hairy elephants but they were not. Detailed knowledge of what they really looked like was only available after the first remains were recovered in the early decades of the 20th century. Only then was it realised that the illustrations made tens of thousands of years ago by our Palaeolithic relatives, who had first hand knowledge of the living beasts, were more accurate than this engraving.

tens of thousands of years old is very remote. This is despite the fact that we now know that high latitudes were being colonized and exploited by modern humans at least 37,000 years ago, even if only on a temporary and seasonal basis for hunting.

It is just remotely possible that some of these ancient people may have been killed in hunting accidents or died from some other cause, but because there would have been relatively few of them in these regions at any time there is only a very slim chance of recovering such a cadaver. In the case of the mammoth, woolly rhino, wolverine and arctic horse, however, they were common in Arctic Siberia during the ice ages and so there is more chance of finding their remains.

As we shall see, the most important of human-related ancient DNA to be recovered has come from the Neanderthal people and that has been obtained from their bones preserved within cool, dark and dry cave environments. First though, let us examine the sort of detailed information that can be recovered from remains preserved in ideal conditions by freeze drying in a glacial environment.

In recent years a number of spectacularly well-preserved bodies have been found in alpine situations. They range from that of the famous British climber George Mallory, who made a fatal attempt to climb Mount Everest in 1924 and whose body was found in 1999 at 8,230m (27,000ft), through the Inuit burials of Greenland, to the oldest of these frozen humans – Ötzi, the Neolithic hunter.

As good as it gets – 'Ötzi', the 5,200-year-old hunter

On 19 September 1991 two German hikers, Erika and Helmut Simon, discovered a frozen cadaver partly embedded in glacier ice at an altitude of some 3,210m (10,500ft) on the mountain border between Austria and Italy. Fortunately the body, soon nicknamed 'Ötzi', the Tyrolean 'iceman', was recovered pretty well intact and still in its frozen state, and has been so preserved ever since. Although emaciated it looked so fresh, however, that it was initially thought suspicious and perhaps a case for the police. They first regarded it as probably being the body of a lost climber, especially as only three weeks earlier the remains of two climbers lost in 1934 had been found.

However, when the clothing and some of the items found with the body were examined more closely it was realized that it had to be much older – 5,200 years old to be precise. The items found included a metal-headed axe and quiver of arrows. Thanks to the excellent preservation, brought about by natural freeze-drying – which co-incidentally reduced Ötzi's body weight by over half from its original 50kg (110lb) to just 20kg (44lb) – it has been possible to carry out some extraordinarily detailed research. This has yielded an unusual amount of information from this almost miraculous find.

Today Ötzi's body is kept at -6°C (21°F) and 99 per cent humidity in a specially-built environmental chamber in the South Tyrol Museum of Archaeology in Bolzano, Italy.

Nicknamed 'Ötzi' after the Ötztal Alps where he was found, this cadaver is one of the best preserved of all prehistoric bodily remains known because it was naturally frozen immediately following his death some 5,200 years ago high above the snowline. The site of his discovery is just on the Italian side of the present day border with Austria. Scientists have been able to recover details such as what he ate for his last meal and where he lived as a child as well as his DNA.

Physically, the iceman was relatively small – just 1.59m (5ft 2.5in) – and, according to bone and teeth studies, about 46 years old when he died – which may have been relatively old for his time. X-rays of his skeleton show that he had an unusual congenital defect – his 12th ribs are missing – and ribs five to nine had been broken and had subsequently healed during his lifetime. There are other bone fractures as well, but these are most likely due to postmortem pressures on the body within the glacial ice that entombed him. There has been a lot of discussion and argument about the reasons for his death and perhaps not surprisingly the media have promoted the more lurid and violent explanations. Nevertheless, there is evidence to show that he did in fact suffer severe trauma before death.

The one fingernail that has been preserved (nails, like hair can easily become detached after death) has three significant disturbances in its growth lines, which suggest that he had been severely ill during the last few months of his life. Examination of his gut has recovered evidence of whipworm parasites, which can cause dysentery, but the extent of his infestation is not known. Tattoos on his lower spine, right knee and ankle are close to acupuncture points and may have been attempts at alleviating some painful condition, but X-rays show no signs of arthritis, which would have been the most likely problem. Other minor conditions include frostbite to the little toe of his left foot, and the presence of human fleas, but they would have been normal complaints for people with his lifestyle.

The most intriguing detail to emerge from a recent X-ray (in 2001) is a stone arrowhead lodged in his back, below the shoulder. It is not yet known what damage it did – whether it severed any arteries for instance – but it would require defrosting of that area of the body and deep probing of the tissues in order to find out and scientists are reluctant to do this. In addition, a deep 3.7cm (1.5in) cut, penetrating to the bone, has been found in his right hand.

Detailed pathological examination of the damaged tissue shows that this cut occurred three to eight days before his death, possibly at the same time as the arrow wound, in which case it would suggest that he survived just a few days before succumbing. There has been a claim that specks of blood from

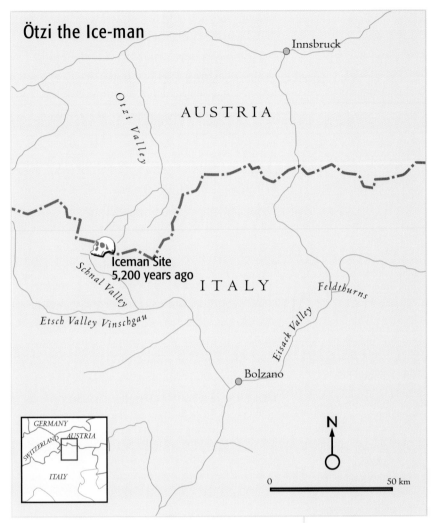

Ötzi the Ice-man

another person have been found on his clothing but that has not been substantiated yet.

In any event, it certainly does look as if Ötzi met a violent end and there is ample evidence to substantiate that.

However, as to exactly in what time of the year he succumbed, the clues conflict...

Some evidence indicates that he may have died in autumn because a sloe (blackthorn) berry was found near the body and in his clothing there were grain kernels, which might have been lodged there during threshing. However, microscopic examination of some digested food residue recovered from his colon shows the presence of well-preserved pollen from the hop hornbeam. This small tree grows only to altitudes of about 1,200m (3,600ft) above sea level and produces copious amounts of pollen in late spring and early summer. Hence, it is likely that as he made his way up into the mountains Ötzi drank stream water upon which the pollen floated. The

Ötzi, the Ice-man. The exceptionally well preserved, frozen remains of a 5,200 year-old man were found in 1991, high in the Tyrolean Ötztal Alps of Italy, close to the Austrian border. Fortunately the cadaver was recovered whilst still frozen, allowing a whole battery of scientific investigations to be carried out. His mitochondrial DNA show that he came from the immediate area. And isotopic data from his teeth suggests that he grew up near Feldthurns in the Eisack Valley but that he later moved to the Vinschgau region of the Etsch Valley before finally moving north to the Schnal Valley, above which he met his death in mysterious circumstances.

Ötzi's clothing and equipment show that he was surprisingly well equipped. Recent testing has revealed his simple footwear to be remarkably effective.

autumn fruiting sloe may well have been carried as dried fruit from the previous year and the grain could also have been embedded in his clothing.

So, for the moment at least, whether Ötzi died in spring or autumn appears to be open to debate, but if we accept the hornbeam pollen as the least problematic piece of evidence, then he probably died in late spring or early summer.

A hardy, self-reliant hunter

A year after the body's discovery, the site was thoroughly combed and several more bits and pieces recovered from the melting ice. Altogether, a surprising amount has been recovered of his clothing and equipment, which show that he was very well kitted out as a hunter and could normally survive in all weathers. He had three layers of clothing with leggings and loincloth overlain with a tunic, made of deer and goatskin and covered with a cape of grass and the tough bark fibres of the linden tree. Finally, there was a bearskin cap and shoes. The latter had a sophisticated construction, which included bearskin soles, goatskin uppers and an insulating layer of grass.

In fact, a modern reconstruction of his shoes has revealed that they are remarkably effective – and comfortable!

His hunting kit comprised a yew-handled axe, 80cm (30in) long with a 9cm (4in) copper blade, a dagger with an ash-wood handle and flint blade, an

Preservation of the footwear, clothing and equipment carried by Ötzi show how well adapted he was to life in his alpine surrounding. He had carefully selected the best local materials and worked them with a considerable degree of sophistication to equip himself well for his lifestyle as itinerant hunter. But he also has wounds that show he may have met a violent death.

unfinished yew bow 1.8m (5ft 11in) long, and a hide quiver with 14 arrows, each 85cm (33in) long. Only two, however, were finished with a flint arrowhead and fletches made of feathers. He also had a belt pouch containing his tinder fire-making kit comprising natural crystals of iron pyrites and flint for making sparks and a tree bracket fungus called the 'true tinder fungus'.

His wood backpack, the remains of which were also found at the site, might have originally carried a net, hide thongs onto which two pieces of birch bracket fungus were threaded and two birch bark containers. One of these could have been used for carrying fire embers (preserved as bits of charcoal) wrapped in leaves of Norway maple. Altogether, these artefacts show that he was clearly an experienced and skilled 'woodsman' who used a large variety of materials from the surrounding environment for his everyday survival.

Sophisticated geochemical analysis of his teeth, bones and gut reveals the presence of certain trace elements and isotopes, which can be compared with those occurring within the surrounding environment, especially in the soil and water. The closest 'fit' suggests that he originated some tens of kilometres further south in Italy in the Eisack Valley near the present village of Feldthurns, but spent his youth slightly further north around the valleys of the Etsch and Schnals and closer to where he eventually died. Overall he probably never moved more than 60km (40 miles) or so from where he was born. However, DNA analysis shows that genetically he came from a typically north central European population rather than from one of Mediterranean origin.

Analyses of the relative proportions of stable isotopes (especially carbon 13 and nitrogen 15) in his bones also give clues as to the sort of food he typically ate. Nitrogen 15 tells us about the amount of animal as compared with plant protein in his diet and it appears that about 30 per cent of his diet was derived from animal protein. Carbon 13 tells us something about the type of plant food and whether seafood was part of the diet.

Not surprisingly, as he was so far from the coast, seafood played no part in his diet, which was predominantly derived from cereal grains supplemented with other seasonal plants especially perhaps fungi, fruits and nuts, some of which may well have been dried and preserved. Analysis of some residues in his gut confirms this, showing that his last meal included some ground cereal grains, wild goat and red deer meat.

The amount and detail of the information obtained from Ötzi is highly unusual. However the same methods of investigation have been used on the remains of our much more ancient ancestors and relatives and have produced some very valuable results.

Ötzi's story shows just what can be achieved when all the techniques of modern science are applied to archaeology and when the remains are particularly well preserved. But the story also tells us something else that is as important today as it was in his time over 5,000 years ago: all life accommodates (or adapts as we should say) to the prevailing environment and climate.

We should also realize that our lives even today are still more intimately related to the prevailing environmental conditions than many of us may think. Despite all our technology, which allows us to travel from the equator to the poles and survive with appropriate shelter, clothing, food and water, we are utterly dependent upon it. If the surrounding climate and environment rapidly changes, unless we have equally adaptable means to feed, shelter and protect ourselves from the elements, our lives may be at risk.

That the climate has changed in the past and will change again in the foreseeable future is an established fact and certainly must have impacted upon our ancestors and relatives. In order to interpret the impact upon them we need to understand the basis for climate and environmental change.

Hard times, changing climates

While we are right to be worried about climate change today and how it might impact upon us and our surrounding environments, the last 10,000 years have enjoyed unusually stable climates compared with the more distant past. We now know that the last 2.5 or three million years have seen numerous and increasingly dramatic swings in climate from phases that were warmer than today into glacial phases that were significantly colder.

The latter we know as the Quaternary ice ages and when they were at their most extreme ice sheets

extended thousands of kilometres from the poles into lower latitudes and mountain glaciers grew even in the tropics. One of the main sources for our present understanding of climate change might seem rather unlikely but the shells of sea-dwelling micro-organisms called foraminiferans have proved invaluable as proxy records of past climate change. Another is derived from the ancient layers of snow and ice buried deep within the thick ice sheets of Greenland and Antarctica (see box on recording climate change below).

The cluster of the continents in the northern hemisphere saw the greatest development of the ice sheets. From the Arctic a huge ice sheet developed to cover most of Canada and beyond the Great Lakes into what is today the United States of America. The glaciation of North America was probably instrumental in delaying the entry of modern humans into the continent. Indeed it is still something of a puzzle as to how and when the first humans did migrate from Siberia into North America. Until recently it has been thought that modern humans could not have entered the continent until the glaciers began to wane and melt away. But there is growing evidence that the first human occupants entered much earlier, perhaps during an interglacial phase.

The more ancient occupation of Eurasia by human relatives as long as two million years ago was

Like Ötzi's, our lives are intimately related to the prevailing environmental conditions; in fact, more than many of us realize today.

Recording climate change

For over 160 years now it has been evident that high latitudes suffered extensive glaciation in the geologically recent past. Evidence from landforms, deposits and fossils all point towards major climate change that oscillated between cold and warm periods. New data from a variety of sources such as cores from the deep-ocean floor, lakebeds and the polar icecaps have all provided detailed and complete logs for significant periods of time within the Ice Ages. From the ice and deep sea records, various proxy measures of climate related change, such as

freshwater input into the oceans (relating to global ice volume) and deep sea temperature has been obtained and related to the recognised timescale. Lake sediments provide fossil pollen and other fossil organisms that are climate sensitive and again changes in their abundance can be timed.

All this data can now be combined with climate models to provide a much better idea of how climate has changed, especially its rate of change. The latter turns out to have been alarmingly rapid at times.

The investigation of Ice Age climates is extremely important for our understanding of how climate will change in the immediate future. Analysis of oxygen isotopes from polar ice-cores and microfossils in deep-ocean sediment cores show that sometimes climate fluctuated over several degrees Celcius within a century or so. We are presently living in a warm interglacial climate.

This diagram indicates the level of sophistication and detail that has been obtained in recent decades about these past climate changes. The top bar covers the whole of the Ice Age (2.6 million years) and lower bars show increasing detail over the most recent intervals down to the last 120,000 years. The peaks on the histogram lines indicate cold stages (even numbers) whilst the valley sand troughs are warm stages (odd numbers), e.g. the last 10,000 years. The polarity chrons represent the periodic reversals of the Earth's magnetic field (see page 120).

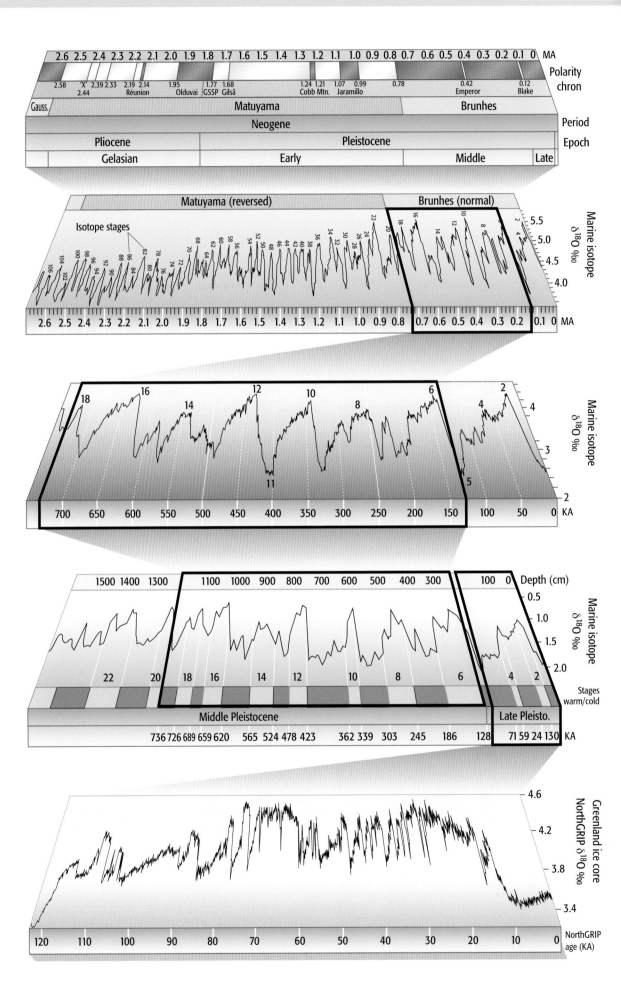

also controlled by fluctuating climate and environmental change.

We now know that the Arctic ice sheets in northern Asia were not nearly as extensive as previously thought, although not so long ago it was generally thought that as in North America, vast ice sheets covered much of northern Asia from the Arctic Ocean down across Siberia.

Such permanently frozen and ice covered landscapes would have made the region virtually uninhabitable but there is contradictory fossil evidence that much of this terrain was in fact inhabited by vast herds of grazing animals of the ice age, such as mammoths, bison, caribou and horses. These animals would have depended upon extensive cold prairie grasslands – and these are not compatible with a significant ice cover. Furthermore, the recovery of fossil pollen from many ice age sites and even the stomach contents of frozen mammoths seemed to confirm the existence of such grasslands that were much more extensive than anything that survives today. Now modern climate modelling shows that the Eurasian ice sheets were largely constrained to Scandinavia, northwestern Europe and the interior mountain ranges of Asia. This was not because the continental interior of Siberia was not cold enough – on the contrary, it was intensely cold – but the air was too dry to generate much snow. In addition, the overall climate was intensely cold and inhospitable during extreme glacial phases. Neither the supposedly cold-adapted Neanderthals, nor our modern human relatives, who had more recently come from Africa, could tolerate the conditions. In some ways it is a wonder that they ventured as far north as they did.

Beyond the ice front there was a vast swathe of frozen ground called permafrost, which was frozen solid to a depth of nearly 100m (300ft). Only the top 10cm (4in) or so thawed during the brief 'summer' weeks, but that was enough to support the growth of tundra plants – especially grasses in the continental interior. These tundra steppe grasslands in turn provided grazing for herds of horses, mammoths, bison and deer along with other more solitary grazers such as the woolly rhinoceros.

Despite the presence of so much 'meat on the hoof' human hunters were not able to avail themselves of it until they had developed hunting technologies and suitable protective clothing and other equipment necessary for survival out on the bitterly cold and exposed steppe. It was thought that the first penetration of such cold northerly latitudes did not happen until somewhere between 14,000 and 13,000 years ago, at the end of the last glacial.

However, there is recent archaeological evidence showing that some hunters did in fact penetrate as far north as the Arctic Circle around 40,000 years ago during the last glacial, although the incursion may well have happened during one of the several brief 'warm' spells. Nevertheless, palaeoclimate indicators show that Mamontovaya Kurya at 66°N, was still bitterly cold with winter temperatures plummeting to -30°C (-22°F) or more.

The mystery of Mamontovaya Kurya

At the Mamontovaya Kurya site near the Polar Ural Mountains, a few stone tools and numerous animal bones have been recovered from beneath 12m (40ft) of sediment in a riverbank section. Most of the bones are those of mammoths and many of them have cut marks typical of butchery with sharp stone tools. There is also a mammoth tusk scored with a series of cut marks suggesting that it was used as an anvil on which other material was cut or chopped. The stone tools include a small bifacial tool and a large scraper typical of the Mousterian (Middle Palaeolithic) or early Upper Palaeolithic age.

'Meat on the hoof' on the tundra steppe grasslands remained out of reach until human hunters developed hunting technologies and suitable protective clothing.

First Arctic hunters – The first record of human relatives penetrating as far north as the Arctic Circle comes from Mamontovaya Kurya, on the south shore of the River Usa in the Polar Urals. During a warm phase, some 37,000 years ago within the last glacial, hunters who may have been either Neanderthals or modern humans, reached used this site as a base for butchering their prey of reindeer, horse and wolf and scavenged older mammoth remains, perhaps for the ivory tusks. The next northernmost site of Byzovaya is much younger (28,000 years old) and is definitely a modern human one.

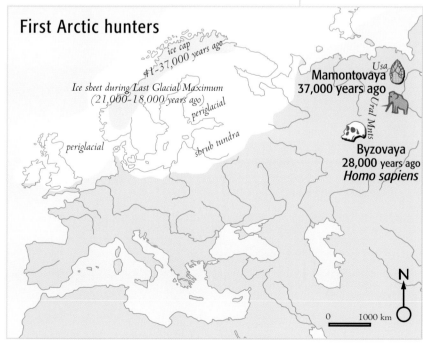

First Arctic hunters

ice cap
41–37,000 years ago

Ice sheet during Last Glacial Maximum (21,000–18,000 years ago)

periglacial

periglacial

shrub tundra

Usa
Mamontovaya
37,000 years ago

Ural Mnts

Byzovaya
28,000 years ago
Homo sapiens

N

0   1000 km

If one can identify plant remains from archaeological sites it should be possible to establish what the climate was like at the time at that particular site.

Fossil plant pollen, especially that from climate sensitive species, such as the birch illustrated here, is used to reconstruct vegetation patterns in the past and their migration in relation to climate change. The pollen is preserved in certain kinds of sediment such as lakebed sediments from which cores may be obtained and the pollen extracted for analysis.

Such tools are characteristic of that period in the development of stone tool technology made by both late Neanderthal populations and early Eurasian modern humans. So it is not yet entirely clear whether the hunters were Neanderthals or modern humans. Unfortunately, no human-related bones have been found that might help determine the matter. If Neanderthal hunters were responsible then we would have to reassess the capabilities and known geographical range of the Neanderthals. Even if the site happened to be just a temporary summer 'stopover', visiting it would have required considerable planning and organization. Normally, such social skills are associated primarily with modern humans and it may well be that it was they who survived a hunting trip to Mamontovaya Kurya.

Until recently it has been very difficult to place this kind of archaeological evidence accurately within a framework of changing climates over time. It has taken over a century for scientists to piece together the very fragmentary evidence for past climates, especially within a well-constrained time frame. The evidence has been gathered from many different sources ranging from ice, deep-sea and lakebed sediment cores to fossil pollen and beetle remains. Most of these provide what is called 'proxy' or indirect measures of temperate and vegetation change.

For many years it has been known that plants and animals can tolerate only certain climate, soil and environmental conditions and that their distribution patterns very roughly follow a latitudinal or height-related zonation with measurable temperature ranges. It follows therefore that if one can identify plant remains from archaeological sites, it should be possible to get a good idea of what the climate was like at the time at that particular site. In addition, most animals are also constrained by climate but some more so than others. This applies to insects in particular, but, unfortunately, few insects fossilize well except for beetle wing cases (elytra) that are remarkably tough and generally adapted as protective cases to the all important and delicate wings. The proteins of the elytra are toughened to resist physical and chemical stresses and consequently they can be commonly preserved in certain terrestrial environments such as lakebed deposits and as a result are important fossil climate indicators.

Likewise plant pollen is adapted to survive the rigours of transport by a variety of vectors from wind and water to animals and prolonged exposure to chemical attack from the atmosphere by desiccation, oxidation and exposure to UV light. And yet it is still viable when it meets up with the right female plant structure. Not surprisingly, pollen often has a high fossilisation potential and can be incredibly abundant in ancient sediments and very useful to scientists trying to reconstruct ancient environments.

For example, at Mamontovaya Kurya the fossil pollen was found in the sediment layers along with the animal bones and stone tools. The pollen has been identified as mainly that of grasses and herbs with some willow that would have grown as scrubby bushes in more sheltered spots. Elsewhere in Europe similar ice age deposits with fossil insect remains show that between 50,000 and 41,000 years ago, the average annual temperature in the region of what is now Holland was around -1°C (30°F). In the coldest months the temperature dropped to at least -10°C (14°F).

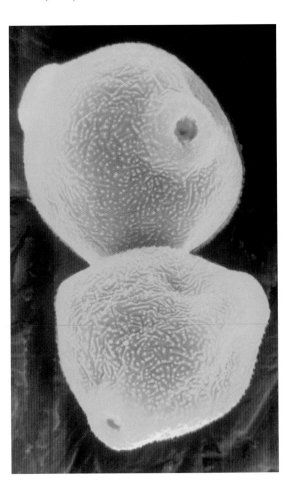

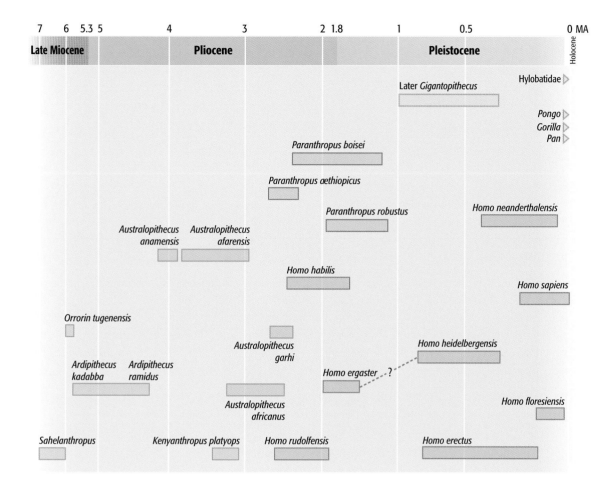

| 7 | 6 | 5.3 | 5 | 4 | 3 | 2 | 1.8 | 1 | 0.5 | 0 MA |

Late Miocene | **Pliocene** | **Pleistocene**

Holocene

Hylobatidae ▷

Later *Gigantopithecus*

Pongo ▷
Gorilla ▷
Pan ▷

Paranthropus boisei

Paranthropus aethiopicus

Paranthropus robustus

Homo neanderthalensis

Australopithecus anamensis *Australopithecus afarensis*

Homo habilis

Homo sapiens

Orrorin tugenensis

Australopithecus garhi

Homo heidelbergensis

Ardipithecus kadabba *Ardipithecus ramidus*

Homo ergaster ---?

Homo floresiensis

Australopithecus africanus

Sahelanthropus *Kenyanthropus platyops* *Homo rudolfensis* *Homo erectus*

The distribution in time of the main extinct human relatives over the last seven million years. Presently there are some 20 fairly close relatives recognised, all of which are extinct except for our own species *Homo sapiens*. Four of these have been recognised within the last decade or so and it is highly likely that several more will be found in the next decade. There is no simple lineage discernable and the overall pattern is that of a shrubby bush rather than a single trunked tree as was once thought.

The colours for the time-range bars show major groupings or related taxa, i.e. members of the genus *Homo* (red), members of the genus *Australopithecus* (orange), the paranthropines, i.e. robust astralopithecines (brown), the primitive hominids whose interrelationships are disputed (green) and other primates (yellow).

Over the last 100 years scientists have painstakingly gathered such fragmentary information about past climates and their changes throughout the ice ages. Much of the data are derived from the northern hemisphere but some lower latitude information is available. From these data a record of some six or seven main glacial phases separated by interglacial phases have been recognized in northwest Europe within the past 2.6 million years. Similar successions have been recognized in Russia, China, North America and even as far away as New Zealand in the southern hemisphere. However it can be very difficult to correlate between the successions in these widely separated regions as much of the correlation is based on fossils, which may be rare or absent. There are also many gaps in these records of land-based deposits and their fossils because successive ice advances tend to obliterate previous deposits.

Evidence from the depths

In addition, the ice core data cover only the last four glacial cycles back to around 400,000 years ago.

However, the past few decades have seen the recovery of sediment cores from the deep ocean. They provide a wealth of data about climate change over the last few million years. Climatologists have known for a long time that there is a closely connected feedback between oceanic and atmospheric circulation. Air temperature and humidity are greatly affected by the state of ocean surface water with which they come in contact.

For instance, northwest coastal Europe has a relatively warm and wet climate that reaches almost up to the Arctic Circle because of the ameliorating effect of the warm Gulf Stream, which flows northwest across the North Atlantic. Consequently, the coast of Norway is ice free whereas on the other side of the Atlantic, sea ice and icebergs are carried much further south (below 50°N) on the cold Labrador Current – as the Titanic famously found to its cost on 15 April, 1912. However, if the North Atlantic circulation pattern were to change and the Gulf Stream ceased to flow northwestwards, northwest Europe would rapidly acquire a climate closer to that of Labrador.

The evidence locked up in the ice cores

Over the last few decades technological advances have made it possible to recover deep ice cores from polar environments such as the Greenland and Antarctic ice sheets and sediment cores from the deep ocean floor. As surprising as it might seem, these unlikely sources are providing some of the most continuous and detailed records of past climate change during the recent ice ages.

Sediment that builds up on the ocean floor contains the shelly remains of ocean-dwelling creatures, especially very abundant micro-organisms called foraminifers. These single-celled (protistan) creatures build shells from minerals dissolved in seawater, especially calcium carbonate ($CaCO_3$). Oxygen (O) is one of the elements present as two different stable isotopes, of which O^{16} is the more common and O^{18} relatively less common. The ratio of these isotopes reflects the concentration of fresh water to saline water in the ocean. Today with relatively high and increasing seawater temperatures the concentration of the heavier O^{18} isotope is low and diminishing.

Conversely, in the depth of an ice age when temperatures are low more freshwater is locked up in ice sheets and glaciers and the relative concentration of O^{18} increases. Thus, by measuring the ratio of the isotopes from shells found at different layers in the sediment sequence it is possible to obtain a proxy measure of temperature change in the oceans. When the history and pattern of these changes are calibrated by radiometric dating the changes can be accurately timed. By means of these methods, some 52 successive couplets of glacial and interglacial ages over the last 2.7 million years have been detected from the ocean record.

The polar ice caps have been accumulating annual increments of snow for several hundred thousand years. The snow is precipitated from atmospheric water vapour that is initially derived from evaporation of ocean waters. Over time successive layers have built up into massive blankets over 3km (2 miles) thick with the older layers below becoming progressively compressed into ice.

When seawater evaporates from the oceans, water molecules with the lighter isotope (H_2O^{16}) evaporate faster and thus enrich atmospheric water vapour with the lighter isotope. Snow from this vapour is therefore relatively enriched in O^{16} and depleted in O^{18}. Consequently, the larger the volume of ice trapped on land in the form of ice sheets and glaciers, the higher the proportion of O^{18} in seawater. During phases of maximum glaciation about 3 per cent of ocean water is removed and 'ponded up' in land-based ice. So higher O^{18} values indicate larger ice caps and lower sea levels.

Retrieval of cores from the ice sheets and analysis of successive related climate changes have showed some extraordinary features. Particularly worrying is the repeated evidence for very rapid climate change in the past. Chronological calibration of these changes shows that annual temperatures have repeatedly changed by as much as 5°C (9°F) within a few decades or a single human lifetime. Such changes must have had a dramatic effect on the environment, the plants, and animals that depended on the plants, including our ancient relatives and ancestors.

The impact of rapid climatic changes

Modelling of ocean circulation patterns and their potential changes, such as 'turning off' the Gulf Stream, suggests that the effect on climate change could indeed be remarkably rapid as indicated by the ice core record. Such rapid fluctuations would have had a drastic effect, especially at the boundaries between vegetation zones. For instance, during the last glacial tundra type grasslands extended southwards over much of Europe and even as far south as Greece and the Mediterranean coast.

Modern climate modelling has become increasingly sophisticated and can now make allowances for many different factors ranging from topography to humidity. A detailed spatial and temporal framework for climate change has been established for the last 60,000 years with a temporal resolution of around 5,000 years within which the archaeological record of bones and stones can be set. This period is particularly interesting as it covers the last 'gasps' of the Neanderthal people and their overlap in time and space with the incoming Cro-Magnon modern humans.

Around 65,000 years ago the world was in the grip of a glacial maximum and even the Neanderthals were pushed south to Spain, southern France, Italy and Greece. Then, as the climate warmed from 60,000 years ago, the ice sheets retreated and the Neanderthals expanded north again up into northern Germany and Russia. By 45,000 years ago their expansion seems to have halted and the number of known Neanderthal

Climate change could indeed be remarkably rapid with drastic effect, especially at the boundaries between vegetation zones.

(Mousterian) sites can be seen to be reducing significantly while the incoming Cro-Magnon modern humans make their first appearance. The latter had spread across the whole region by 40,000 years ago and were occupying a similar density of (Aurignacian) sites as the Neanderthals. By 35,000 years ago, as another glacial takes hold, the number of Neanderthal sites is reduced and they seem to be retreating into their ancient strongholds in Spain and the south of France. However, although the number of Aurignacian Cro-Magnon sites is reduced, they still have more northerly and easterly outposts than the Neanderthals.

A new feature was the appearance of a new wave of modern humans characterized by the somewhat more advanced Gravettian cultural artefacts. They soon expanded across the whole region and show less sign of retrenchment in the face of worsening climate than either the Neanderthals or the Aurignacians. By 30,000 years ago and the depths of

the last glacial period, the Neanderthals have been reduced to just three locations in Atlantic southern Portugal, the south of France and the Ardennes. By comparison the Aurignacian Cro-Magnons and Gravettians were still scattered over a wide region from Britain to Russia and south to Italy and Iberia. The Neanderthals did not survive this last retreat and even the Aurignacian Cro-Magnons were severely depleted by 25,000 years ago, after the demise of the Neanderthals.

The only group to persist in any significant way was the Gravettians, presumably because of their advanced technology and social structures. The Neanderthals may have been cold adapted to some extent but they were still not able to cope with extreme glacial conditions and perhaps the more successful incoming modern humans. Even that first wave of incomers, the Aurignacians, seems to have been eventually defeated by deteriorating climate; only the Gravettians could cope.

Different materials for different uses

While our ancestors found stone to be a remarkable material that could perform a great variety of jobs depending upon the exact properties of the particular rock or mineral selected, it also has some severe limitations mainly because of its brittleness. By contrast other naturally occurring materials such as wood, bone, ivory and antler are less brittle and although not so hard and sharp as stone have other complementary properties. They can be more easily carved, sharpened, bored and bent and are not so heavy.

However, there is a problem for archaeologists in knowing when these basically organic materials were first used by our ancestors because their preservational potential is much lower than that of stone especially within the African

context. Nevertheless, there is plenty of indirect evidence for the use of bone and antler and direct evidence from Africa that is of considerable antiquity.

The production of relatively sophisticated stone tools, such as the Acheulian hand axe, and many other ancient tools require delicate flaking and reworking. Experimentation has shown that such manufacture probably required the use of so-called 'soft hammers', especially those made of bone or antler, rather than 'hard' stone ones. In Africa the production of such carefully worked blades dates back at least 240,000 years ago but we do not know which human related species made them. The oldest of such indirect evidence was found in central Kenya in the same strata that have

yielded the first fossil chimps (the Kapthurin Formation).

The oldest direct evidence for the use of bone comes from a series of remarkable barbed harpoon-like points from Katanda in Zaire. Unfortunately they have been difficult to date but ESR techniques suggest that they are between 160,000 and 90,000 years old and their manufacture is almost certainly linked to our species.

In Europe there is a clear correlation between the appearance of bone, antler and ivory in the archaeological record and the tool culture associated with the Upper Palaeolithic linked to the arrival of modern humans around 32,000 years ago. While the incumbent Neanderthals must have used soft antler and bone hammers in the preparation of

some of their stone tools, they do not have appeared to have made more sophisticated tools from these materials. In contrast, modern humans eventually developed the necessary technologies for the manufacture of bone needles, which they used for sewing their clothing, and barbed harpoons for fishing.

The curious thing is that our ancestors and distant relatives were familiar with many of the properties of bone, antler and ivory. Over a period of two million years stone tools were used to butcher animal carcasses for meat and to break open bones for marrow. And yet, it appears to have taken an inordinate length of time for the full potential of these materials to be fully recognised.

CHAPTER THREE

Our view of our antiquity and how it has changed

Today the scientific world-view of human origins and antiquity embraces evidence from a wide range of sources – from fossil bones and archaeological remains to biomolecular and genetic data. All of these diverse sources unequivocally point to our evolution as primate mammals from an ancestor shared with the higher apes some seven million years ago. This ancestor lived, evolved and diversified in Africa into at least 20 different species, a few of which have dispersed beyond Africa. All of these species have now become extinct bar one – *Homo sapiens* – and some of these extinctions have been very recent – within the last few tens of thousands of years.

Overlays of painting by different hands are clearly evident
in this Palaeolithic cave painting.

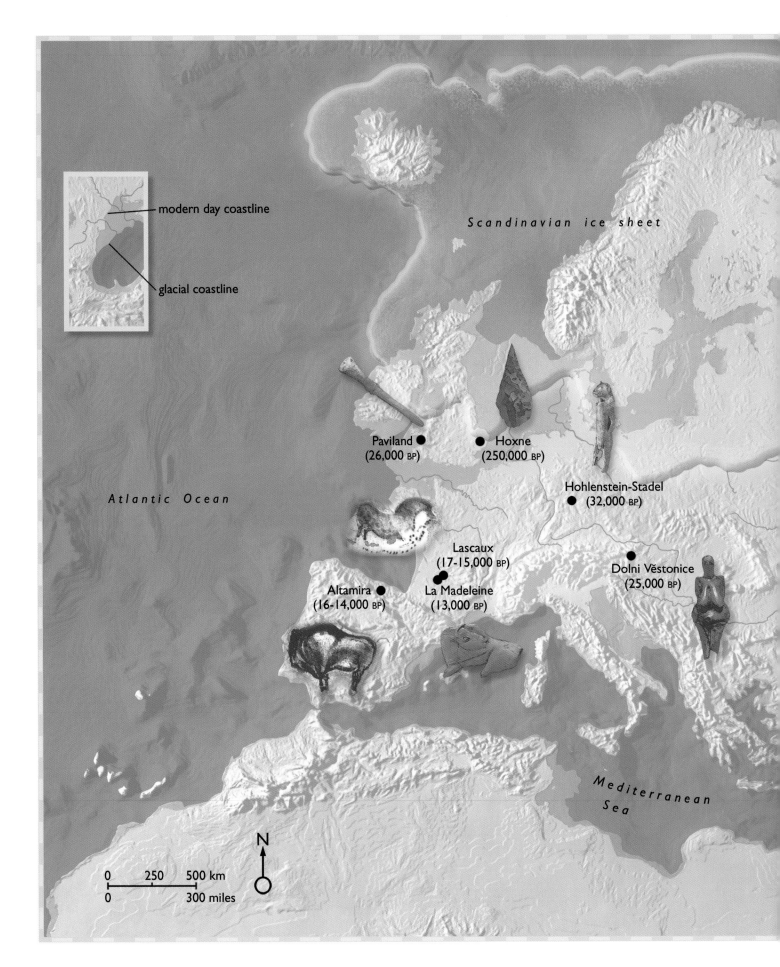

modern day coastline

glacial coastline

Scandinavian ice sheet

Atlantic Ocean

Paviland ●
(26,000 BP)

● Hoxne
(250,000 BP)

Hohlenstein-Stadel
● (32,000 BP)

Lascaux
(17-15,000 BP)

Dolní Věstonice
(25,000 BP)

Altamira ●
(16-14,000 BP)

La Madeleine
(13,000 BP)

*Mediterranean
Sea*

N

0	250	500 km
0		300 miles

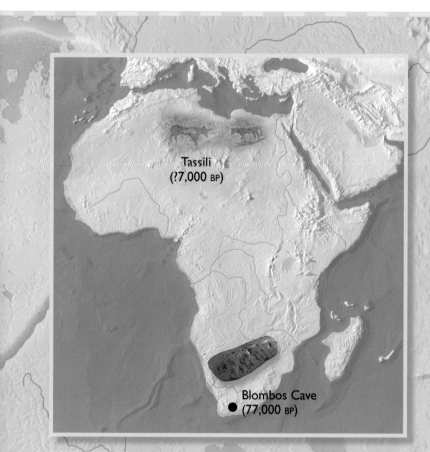

Tassili
(?7,000 BP)

Blombos Cave
(77,000 BP)

Bradshaw

Carnarvon Gorge
(3,500 BP)

Ancient artworks reveal the antiquity of humanity

Clear evidence for the antiquity of humans has been revealed over the last 200 years ever since 1799 when the first manmade flint handaxe was found at Hoxne in eastern England. Successive finds of sophisticated works of art associated with the remains of extinct Ice Age animals reinforced the view that our human ancestry extended back many thousands of years into the Ice Age. Cultural artifacts have now been found in southern Africa that date back to around 80,000 years ago but they are all made by our species – *Homo sapiens*.

Paviland – around 26,000 years ago mammoth ivory sticks were placed alongside a modern human burial and covered with red ochre.

Hohlenstein-Stadel – a lion's head has been carved on top of a human body in this 30cm (12in) 32,000-year-old ivory statuette from Germany.

Lascaux – amongst the most famous cave paintings, these relatively recent depictions of bison and rhino, horse and humans, are dated to between 17,000 and 15,000 years old.

La Madeleine rockshelter – this exquisite antler carving of a bison licking its flank is around 13,000 years old.

Altamira – the famous bison images have been dated to between 16,000 and 14,000 years old.

Dolní Věstonice – one of many 'Venus' figurines found scattered over Europe, this one is 11cm (4in) high and was moulded from clay and ash before being fired. It is dated at around 25,000 years old.

Hoxne – found in the 1790s, this handaxe was the first artefact to be recognised as a man-made tool and evidence for the existence of humans alongside extinct animals.

Tassili – the animals illustrated in the rock carvings of North Africa show that the region was previously much wetter with lakes, rivers and woodland.

Blombos – a 77,000 year old red ochre crayon with an engraved pattern is the oldest known 'art work' associated with *Homo sapiens*.

Bradshaw – in the 1890s explorer Joseph Bradshaw discovered some exquisite rock paintings of unknown age in the Kimberley region.

Carnarvon Gorge – hundreds of stencils of boomerangs, axes, hands and even full humans along with engravings of nets have been found in Carnarvon Gorge in Queensland.

The origination of our species happened in Africa within the last 300,000 years or so and by 100,000 years ago modern humans had spread both within Africa and beyond the continent into the Middle East. Over the next 50,000 years our species spread globally to a far greater extent than any previous human relatives. By 50,000 years ago modern humans reached Australia and by around 20,000 years ago they had penetrated eastern Siberia, from where they spread into the Americas. The final phase of the global diaspora took our ancestors out across the Pacific and to New Zealand by around 1,000AD.

This whole picture has been developed only within the last 250 years. For even though there was mounting evidence for human antiquity by the mid 18th century, very few of the natural philosophers of the day (the word scientist was not widely used until the 19th century) would risk openly voicing ideas that were contradictory to the prevailing world-view.

Even in the western world, where modern science was developing, the prevailing orthodoxy of Judeo-Christianity was extremely powerful and all pervasive. Although modern science had grown from ancient and diverse roots which are derived from many different cultures, including pre-Christian ones of the classical world, those of India and the Muslim world, the modern development of science dates from the Renaissance in Europe where Christianity held sway.

A time when 'heretical' opinions could be fatal

It is perhaps hard for us today to realize or understand the power and influence that Christian orthodoxy held over much of western society 250 years ago. As recently as 1619 the Italian philosopher Lucilio Vanini was burned at the stake in Toulouse in France for the atheistical heresy of suggesting that humans might be descended from the apes. Many 18th century naturalists still openly subscribed to the Judaic Old Testament version of the Earth's origin, the special Creation of 'Man' and the story of the Mosaic Flood.

In the first decades of the 19th century, many prominent geologists such as the Reverend Dr William Buckland of Oxford University still

believed in the Creation and the Flood. But then Buckland was also Anglican Dean of Westminster and so was expected to espouse the church's view.

Before science became professionalized many of the other natural philosophers of the late 18th and early 19th centuries were also clerics, especially in Britain where the sciences of geology, palaeontology and biology were developing very rapidly. In order to be appointed to the position of don, those of the universities of Oxford, Cambridge and Trinity College, Dublin, had to be ordained members of the established protestant church and so it is not surprising that so many of them were deeply reluctant to criticize the received opinion of current theology.

Geology confirms Old Testament story?

A group of early 19th century so-called theological geologists, including the Rev Dr Buckland hoped that the emerging science of geology and palaeontology would actually support the general view of prehistory as presented in the Old Testament. They were particularly concerned with the story of the Flood and the special Creation of Mankind. By an extraordinary accident of recent geological history, evidence was emerging in the first decades of the 19th century geological that did seem to support the Flood story, and since human-related fossils were not turning up in the fossil record, the Creation story might be upheld as well.

The story of the Mosaic Flood is a very powerful one and since it was intermingled with actual geographical and dynastic detail of the early history of the Israelites it did seem remarkably convincing. For example, across northern Europe, Asia and the Americas, gigantic fossil bones had been turning up every now and again for many centuries. Often they were found washed out of riverbanks, coastal cliffs or in shallow diggings for sand or clay. For instance, in the 12th century the English chronicler, Ralph of Coggeshall in Essex, recorded in 1171 how the collapse of a riverbank had revealed the gigantic limb bones of a man who 'must have been fifty feet high'.

The bones in question were almost certainly those of an elephant or mammoth, and they are occasionally still found in Quaternary age deposits in this region. A similar thighbone found in 1443 by

There was, understandably, a reluctance to cross swords with the orthodox view of creation when doing so could be dangerous – and in earlier times had been fatal.

workmen digging the foundations of St Stephen's Cathedral in Vienna was again thought to be that of a giant and for centuries was chained to the cathedral door, which was known as the Giant's Door. By the 18th century such bones were no longer seen as those of giants but recognized for what they were – the remains of fossil elephants and other mammals – but they nevertheless seemed very exotic in the European and North American context.

'Mammuts' described

Similarly, stories had been circulating in Europe of discoveries of strange elephant-like skeletons and even corpses in Arctic Siberia, and the circulation throughout Europe of considerable quantities of ivory from the region seemed to confirm the stories. Early scientific expeditions reinforced these views and by the mid 18th century skeletons, tusks and even bits of skin and hair of what became known as the 'mammut' were recovered from the frozen permafrost of Siberia. They were described and illustrated by naturalists such as the German Daniel Messerschmidt (1685-1735) in the Philosophical Transactions of the Royal Society of London, one of the oldest scientific societies in the world.

Even as early as the beginning of the 18th century the society's journal contained a description by the polymathic American scholar Cotton Mather (1663-1728) of a giant tooth '...almost thirteen inches in circumference...' which 'weigh'd two pounds four ounces Troy weight'. This elephant related tooth and other bones had been found in 1705 in the banks of the Hudson River at Claverack, near New York. Scholars from the Americans Benjamin Franklin (1706-90) and President Thomas Jefferson (1743-1826) to the Frenchmen Georges-Louis Leclerc, Comte de Buffon (1707-1788) and Georges Cuvier (1769-1832) were greatly interested in this curious distribution of elephant-related fossils and questioned how it might have happened.

Historically it was just possible that such bones might be the remains of elephants brought to France by the Carthaginian Hannibal for his crossing of the Alps in 218BC. And Emperor Claudius brought elephants to England in AD43, so it might just be possible that the bones were those of these Indian elephants that were being used as 'warhorses'. But there was no way that the Romans could have been responsible for the elephant remains in North America. For the theologically minded, which included most of the 18th century naturalists, the most obvious answer was the Mosaic Flood which had swept the remains of these hot country animals from the tropics into high latitudes where they were deposited along with the bones of many other exotic animals, such as hippos, hyenas and rhinoceros.

The Flood was also invoked as being responsible for the deposition of shells and bones of sea-dwelling creatures far inland and even on mountaintops. But geologists were finding that the known thicknesses of fossiliferous strata

The Romans could not have been responsible for the elephant remains in North America.

The 18th century naturalist Johann Scheuchzer produced an illustrated bible that included contemporary scientific knowledge, here for instance foetal development frames the creation story.

cumulatively amounted to many kilometres and it was hard to explain all this by the action of a single universal flood event. A possible way around the problem was to consider that there had in fact been many floods but by the end of the 18th century many geologists knew that the evidence on the ground did not support a literal interpretation of the Flood story. However there were those such as Buckland and even later the Scots fundamentalist Hugh Miller (1802-56) who were still trying desperately to somehow shoehorn the 'Testimony of

the Rocks', as Miller called it, into a version of the Flood story.

Even as late as 1930, the well known English archaeologist and writer on archaeology, Harold Peake, felt obliged to write a book entitled *The Flood: New Light on an Old Story*. He wanted to explain in layman's terms how the then modern archaeological discoveries at Ur and Kish in Mesopotamia (today's Iraq) impacted upon the Old Testament story. Excavations had revealed that there certainly were major floods around 6,000 years ago in the region and these were indeed catastrophic for the inhabitants – but they were **not** universal and predated the Old Testament flood story.

Similar major flooding occurred throughout much of central and southern Asia, northern Europe and central North America following the melting of the vast ice sheets and glaciers of the last ice age from 10,000 years ago. Huge meltwater lakes were formed in places and these often burst through their natural dams to flood enormous areas very rapidly. Any living creature in their path would have been swept away, apart from birds that could have flown away. Hence, it is not surprising that many different peoples have flood stories in their folk memories and tales.

Victims of the Flood?

One of the really interesting aspects of the Flood story was how it related to early human history. According to the Old Testament account there should have been many human sinners who were drowned along with all those animals and plants that could not be squeezed into the Ark. And yet they were not evident as fossils. Occasionally, however, fossils were found and were hailed as victims of the Flood.

Most famous was a skeleton found in strata quarried for lithographic stone near Oeningen and Lake Constance in Switzerland in the early 18th century. Johann Jakob Scheuchzer (1672-1733), the most famous naturalist of his day, whose publications were widely distributed and known throughout Europe and beyond, obtained the relic. Scheuchzer had already described a number of fossil plants as remnants of the Flood in his book *Herbarium of the Deluge*, published in 1709.

Scheuchzer's depiction of Noah's Ark prior to the Flood is supported by illustrations of fossils (insects at the top and a seally at the bottom of the frame) that were generally considered at the time to be evidence for that Flood.

Now he delighted in being the first to describe and picture this fossil as *Homo diluvii testis* – the 'man who was a witness to the Flood'. The skeleton did indeed show a flattened helmet-shaped skull, long backbone and four limbs splayed out to the side. Although Scheuchzer was a qualified physician and should perhaps have known better, he was presumably so keen to believe that he really did have

the skeleton of a Flood victim that he suspended his anatomical judgment. In his defence, many other experts were convinced of the veracity of the find – until 1809 when Georges Cuvier made a public display of debunking it in front of an audience in the Tyler Museum in Haarlem where the specimen was kept after Scheuchzer's death.

Cuvier displayed his considerable expertise as a comparative anatomist by predicting the existence of a distinctive amphibian bone on the specimen – and then uncovering it to show that the skeleton was *actually* that of a giant salamander. Cuvier did not believe that any human-related fossils would be found because he still believed in special creation for humans.

Missed opportunities

With the benefit of hindsight we can now see that there were a number of other discoveries, made from the latter part of the 18th century onwards, which were either misinterpreted or whose importance was not recognized. For instance, in the late 1790s a strangely fashioned flint was discovered by workmen digging out sand and gravel at Hoxne, Suffolk in the east of England. They must have been curious about its unusual shape or were hoping for a reward because they passed it on to a local antiquarian John

Johann Scheuchzer, like most of his contemporaries was convinced that the study of natural history would reveal the truth of the Christian deity's wisdom and creative design of the natural world.

Scheuchzer suspends judgment – and his silent witness turns out to be a giant salamander.

This fossil skeleton was seen by Scheuchzer as the remains of a drowned sinner who was witness of the Noachian Flood. Named *Homo diluvii testis*, the fossil was eventually recognised as that of a giant salamander by the French anatomist Georges Cuvier in the early 19th century.

Frere (1740-1807) whom they knew was interested in such things. A Cambridge graduate, fellow of Caius College, high sheriff of Norfolk, member of Parliament and Fellow of the Royal Society, Frere was a typical member of the intelligentsia of the day with catholic interests in a wide range of subjects.

He recognized that as the flint was so carefully fashioned into a pointed pear-shape, with flattened sides and sharp edges, it could only have been fashioned by human hands rather than by the natural processes of weathering and erosion. He also knew that it had been dug out from sands and gravels that were several feet below the surface.

It was from exactly such deposits that the bones of extinct mammals were found and so Frere concluded that this flint hand-axe must have been 'fabricated and used by a people who had not the use of metals...' and that 'the situation at which these weapons were found may tempt us to refer them to a very remote period indeed, even beyond that of the present world'. The manufacturers of this flint hand-axe must have coexisted with the extinct mammals whose skeletal remains were being found increasingly in superficial deposits. At the time the general interpretation was still that these were creatures that had been drowned in the Flood.

Frere published a description, including a careful illustration of the axe and his conclusions about its origin in the journal of the Society of Antiquaries in 1800. His remarkable insight seems to have been the first to be made, but its significance was apparently overlooked at the time. Two decades later William Buckland missed *his* chance to make an even more startling discovery because his theological mindset completely blinkered his 'vision'.

In the 1820s Buckland's attention was drawn to a limestone cave called Goat's Hole at Paviland on Gower Peninsula that extends from the South Wales coast out into the Bristol Channel. Buckland had realized that cave deposits often contained very interesting fossils within the accumulated sediment deposits that formed the cave floor. His excavation at Paviland was one of the first reasonably careful scientific investigations of cave floor sediments in which the position of the various layers was noted along with the distribution of bones and artefacts within the sediments. Some 5,000 artefacts were recovered.

Most importantly, Buckland found a shallow grave containing a skeleton along with seashells perforated for personal adornment as perhaps some sort of necklace. There were also carved pieces of

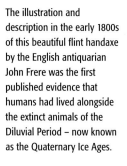

Buckland missed the chance to describe one of the richest and best Palaeolithic sites in Britain – and his 'she' was a 'he'.

The illustration and description in the early 1800s of this beautiful flint handaxe by the English antiquarian John Frere was the first published evidence that humans had lived alongside the extinct animals of the Diluvial Period – now known as the Quaternary Ice Ages.

T.R. Underwood, del. 1797.

Flint Weapon found at Hoxne in Suffolk.
Archaeologia Vol. XIII. p.204.

mammoth ivory, including bracelets and long thin wands. Everything had been dusted over with a scattering of red ochre.

It was evident that the body had been buried with some ceremony and Buckland, from his examination of the bones, concluded that the body was that of a young woman who he nicknamed the 'Red Lady of Paviland'. Despite the occurrence of stone tools, mammoth ivory and the bones of extinct mammals, Buckland could not face up to the simple truth of the matter and its implication – that this person had lived and died alongside those extinct creatures.

Instead he had to produce the convoluted explanation that she was a Welsh tribeswoman who lived in Romano-Celtic Britain. He reckoned that upon her death her kinsfolk buried her in the cave where they found the mammoth ivory, which they then carved and placed in the grave.

In taking this route, Buckland missed the chance to describe one of the richest and best Palaeolithic sites in Britain.

We now know that 'she' was a 'he', about 25 years old and stood some 1.7m (5ft 6in) tall. Radiocarbon dating has produced a date of around 26,350 years for the skeleton and it turns out that he was one of the anatomically modern Cro-Magnon people who migrated into Western Europe around 40,000 years ago.

Ice Ages replace the Flood

During the first few decades of the 19th century European geologists noticed that many of the strange geological phenomena associated with Alpine glaciation resembled features that had previously been interpreted as being caused by the Flood.

There was a growing understanding of the processes and deposits of 'rivers' of ice flowing out from the mountains and down far onto the lowlands under the influence of gravity. There was much argument, but eventually the glacialists won the day. For instance, continental experts such as the Swiss geologist Louis Agassiz (1807-73) were able to persuade even the most committed theological geologists like Buckland that there were plenty of signs of the erosive power of glaciation in upland Britain. In addition, there were glacially derived deposits plastered over the landscapes of much of lowland Britain, Europe and North America extending for many miles beyond the maximum reach of the ice sheets and glaciers. This acceptance of the impact of glaciation on the land led to the story of the Flood story being replaced by an almost as dramatic series of Ice Ages.

By this time it was also realized that abundant wildlife had occupied the landscapes beyond the ice and that these animals included a great diversity of cold-adapted mammals including vast herds of mammoths, bison, horses as well as woolly rhinoceros and big predatory cats. Furthermore, during glacial phases of lowered sea levels the more mobile species were able to migrate from continental mainlands to offshore islands and even from continent to continent. For instance animals migrated from Europe into the British Isles and small bands of human hunters who depended on the animals for food, clothing and many other necessities of life followed them.

Today the Goat's Hole cave at Paviland is on the coast, but was over a kilometre inland when the sea level was lower. During such a time it would have overlooked a wide, flat, low-lying coastal plain that stretched from the shoreline back to low-lying limestone cliffs. It therefore would have been an excellent vantage point for viewing the movements of game, which during warm interglacial periods consisted of wild cattle (aurochs), horses, deer, elephants, hyenas and so on.

The Reverend William Buckland (1784-1856) was reader in Mineralogy in the University of Oxford, President of the Geological Society and became Dean of Westminster. He tried to combine the emerging data from the geological sciences with his profound Christian beliefs but came to realise that literal interpretations of the Bible could not be reconciled with the geological and fossil record.

During ice ages sea levels dropped, and more mobile animals were able to migrate to the British Isles... and hunters followed them.

Today, the caves in the limestone cliffs of the Gower Coast in South Wales are right by the sea but when the caves were first used by our Palaeolithic ancestors over 20,000 years ago, sealevels were much lower. Then the caves overlooked a wide coastal plain and gave an excellent vantage point for viewing the game that our ancestors hunted for food and other essentials of life in the Ice Ages.

A bone spatula from the cave has been dated at around 23,670 years and the range of dates in respect of various other artefacts found in the cave suggests that it was intermittently occupied for a considerable period.

This is confirmed by climate records - around 23,000 years ago the climate began to descend into another ice age. Ice sheets and glaciers would have progressively penetrated further south from North and Central Wales as the ice age deepened and cold-adapted species such as bison, reindeer, mammoths and occasional woolly rhinos and wolves would have replaced the interglacial animals.

As the ice approached South Wales, the increasingly colder conditions would have forced the cave's inhabitants to abandon it and seek a warmer, more tolerable environment until the climate again began to improve around 13,000 years ago and the beginning of the modern interglacial phase of Holocene times (from 11,000 years ago).

'Man's' place in Nature

By the early decades of the 18th century, naturalists had begun trying to formalize the bewildering array of life according to some 'natural order' by naming and classifying all the different kinds of plants and animals then known. They used Latin, the international language of scholars, and attempted to group like with like in a hierarchy similar to that outlined in the Old Testament story of creation. Carl Linnaeus (1707-78), a very talented Swedish botanist, was the first to give the scheme a scientific basis. The roots of the scheme are much older but today taxonomists recognize the tenth (1753) edition of Linnaeus' book *Systema Naturae* as the formal starting point for the Linnaean system classification as it is called.

The whole basis of the scheme is the recognition of the species as the fundamental unit of biological classification. The name is denoted by a Latin binomial (literally two names), which is made up of the species name (for us that is *sapiens*) preceded by a genus name (again for us that is *Homo*). It was Linnaeus who gave us this technical name meaning 'knowledgeable man'. He even selected himself – a white Scandinavian – as typical of the whole species. While basically there can only be one group of interbreeding individuals, past and present, which can make up a species, there can be more than one species in a genus.

At the beginning of the 19th century, once Cuvier had dismissed Scheuchzer's *Homo diluvii testis* as a fossil salamander rather than a fossil human victim of the Flood, the genus *Homo* was thought to contain only one species *sapiens*. But of course it was also widely recognized that many genera of plants and animals contained many closely related species. When Linnaeus first developed his classificatory scheme in the 1730s, he was attempting to bring together, sort out and pigeon-hole all known forms of life on the basis of their degree of similarity to one another.

The measure of togetherness was fundamentally one of morphological appearances of the constituent parts of organisms. Most of the organisms Linnaeus was dealing with were living creatures but he did include some fossils (mostly fossil plants and shells). Even so, there were no evolutionary implications to the scheme. Linnaeus considered that by developing the classification he was merely revealing the wisdom and order of God's creative powers and grand design for life on Earth. The title page of his *Systema Naturae* has a quote from the Old Testament Psalms (104, verse 24):

> 'O Jehovah! Countless are the things you have made!
> Thou hast made all by thy wisdom!
> And the earth is full of thy creatures!'

A belief in species by design

Like most of the naturalists of the day, Linnaeus firmly believed in the fixity of species... that the deity individually designed each species for a particular purpose in a particular place. However, Linnaeus also recognized that the grand design ascended from the lowest to the highest – and guess who was placed at the pinnacle of earthly life? Even though the species he placed within a single genus could be remarkably similar, that affinity did not imply an evolutionary connection.

Linnaeus also tried to explain the apparently strange way in which different kinds of animals and plants occupied different parts of the world. He was not convinced by the literal interpretation of the Flood but imagined that originally the deity had created life on a mountainous tropical island surrounded by a primeval ocean. The height of the mountain produced a succession of climates from hot tropical at sea level (and home to life forms such as palms and monkeys) to cold at the top, which was home to such life forms as reindeer and lichens. Thus the mountain was a miniature version of the earth and when the waters of the primeval 'Flood' ocean subsided the survivors spread out into the regions that best suited them.

By the 10th edition of his work it had grown to accommodate the totality of living organisms known to Linnaeus from all around the world. The total was a mere 12,000 or so species comprising 7,700 plant species and 4200 animals. However, he was well aware that there were many more to be found, especially in tropical lands, and he encouraged his students such as Daniel Solander (1733-82) to travel the world looking for them.

Linnaeus was not convinced by the literal interpretation of the Flood.

The 18th century Swedish taxonomist, Carl Linnaeus, published systematic works that attempted to name and describe all then known, plants and animals (including fossils). His system of taxonomic nomenclature has been taken as the basis for all subsequent naming of plants and animals.

Linnaeus recognized the class Mammalia and one of the major groups he distinguished within it was the order Anthropomorpha (meaning 'human form') which he later changed to Primates, meaning 'first'. In this order Linnaeus included both *Homo sapiens* and *Satyrus tulpii*, meaning 'Tulp's Ape' (Nicolaas Tulp being a famous Dutch anatomist who dissected a chimpanzee and who was painted by Rembrandt) by which he was referring to the chimpanzees. Their species name was not strictly clarified, however, until Cuvier renamed them as *Simia troglodytes*, although the genus name was subsequently changed again to *Pan*.

A challenge

Other scholars of the day were critical of the inclusion of humans in the same order as the chimpanzees but Linnaeus challenged them to find any anatomical feature that warranted a clearer separation.

Linnaeus knew that he could invoke the posthumous support of a scientific monograph, published by the Royal society of London in 1699, describing the detailed anatomy and skeletal structure of the chimpanzee. The monograph's author was a brilliant English anatomist and physician, Edward Tyson (1650-1708). A young chimpanzee had been transported to England in 1698 as a collector's item of curiosity but it was sick and died soon after its arrival so Tyson had the opportunity to dissect the corpse. Chimpanzees were virtually unknown in Europe in those days because there was little contact with the parts of Africa from which they came; by comparison the orang-utans from the Dutch East Indies were slightly better known.

Tyson's description and illustrations were masterly. The chimpanzee's skeleton still survives in London's Natural History Museum and it is hard to find fault with the published engraving, although he portrayed it standing upright like a human. Tyson enumerated features that compared most closely with those seen in humans (48) and in monkeys (27). He concluded that the animal was therefore an 'intermediate link' between monkeys and humans and by far the closest anatomically to humans. In doing so he was placing the chimpanzees on the progressive scale that dates back to Aristotle and is known as the 'Great Chain of Being' or ladder of nature.

Cuvier, despite his anatomical skills, was mistaken.

According to Tyson, he saw it as a graduated scale running from the organized form of minerals through plants and animals to humans and on through the angels to the deity. For Tyson, a knowledge and understanding of this hierarchy was a matter of respect for the Creator.

The first fossil primate to be discovered was found during the first decade of the 19th century in gypsum quarries in the village of Monmartre, which was then just outside Paris. Monmartre's calcareous strata of the Eocene age were extensively quarried for the carbonate mineral used as the basis of the widely used 'plaster of Paris' and occasionally the quarrymen found fossils within the ancient sediments. They sold the fossils to any interested 'savants' and as professor at the Natural History Museum in Paris, Georges Cuvier was well placed to take advantage of this local supply of remarkable fossils. They included the tiny primitive horse *Palaeotherium*, the first fossil marsupial to be described and the first primate, which he called *Adapis parisiensis* (1821). The genus name *Adapis* means 'towards Apis', the bull god of ancient Egypt. Cuvier misguidedly chose it because of his mistaken ideas about its affinities.

Again with the benefit of hindsight we can see that even with his anatomical skills, the great French anatomist got this one wrong. The specimen was a severely flattened skull and lower jaw and he concluded that it was a primitive pachyderm, an old grouping which included a wide range of animals from tapirs to rhinos, horses, elephants and so on. Cuvier knew that the Monmartre fossils were found in hard lithified sedimentary strata which lay below the loose unconsolidated sands and gravels that contained the remains of the more modern looking mammals of the 'Deluge'. The gypsum beds were therefore older and as he said in 1800 'the older the beds in which these bones are found, the more they differ from those of animals that we know today'.

The first human relatives to be recognized

One of the reasons for the failure to recognize the remains of our ancient relatives and ancestors is that in Europe at least, the remains, such as Buckland's Paviland skeleton, are mostly no more than a few tens of thousands of years old and thus belong to anatomically modern humans. When such remains were found they would have been dismissed as merely those of ancient heathens and some of them were reburied in Christian graveyards.

However, discoveries were eventually made that were distinct enough to raise serious questions about their origin and history, and the most important of these finds was made near Düsseldorf in Germany.

The Neander Valley (Neanderthal in German) was named after the 17th century composer and priest Joachim Neander, who loved its picturesque charm, with the Dussel River flowing between wooded limestone hills. Over millions of years, water has eroded the limestone, creating numerous caves and passages and these added an air of mystery and charm to the valley. By the 19th century, however, growing industrial pressure had led to the widespread exploitation of good quality pure limestone for smelting as well as for lime, and despite the attractions of the Feldhofer Grotto, by the 1850s its commercial potential led to it being quarried.

Then, in August 1856, as workmen dug out the sedimentary deposits which had infilled the cave floor, they came upon some buried bones. Although the occurrence of animal bones in caves was not all that unusual, *these* bones drew the workers' attention and fortunately someone knew that the local schoolmaster Johann Fuhlrott (1803-77) was interested in such things.

Fuhlrott was a natural historian and well read in a wide range of contemporary scientific literature and the bones from Feldhofer fascinated him. He was duly presented with a very well preserved, thick upper part of a skull – the skullcap or cranium plus limb bones and a few other parts of the skeleton. Although he attempted to find out if any other bones were present, he was too late as all the deposits had been dug out and disposed of. Nevertheless, what had been salvaged was significant and he was forcibly struck by the form of the skullcap and the way that it differed from that of modern humans.

Prominent features

The most prominent feature was the prominent thick double arched bony brow-ridge behind which the skull roof sloped gently back so that there was no evident forehead. Also, the sides of the skull were strongly 'pinched' in behind the brow-ridge before expanding again so that from above the brow-ridge looks like a curious visor with prominent side extensions. Fuhlrott remembered that he had seen a skull of similar shape in comparative illustrations of chimpanzee and orang-utan skeletons and skulls published by the prominent English anatomist Richard Owen (1804-1892) in 1835.

Fuhlrott thought the bones important enough to be properly described but knew that he needed a collaborator with the right academic credentials and expertise to do so.

So he took the bones to a professor of anatomy at the University of Bonn, Hermann Schaaffhausen (1816-93), who was equally intrigued. Together they

Finding animal bones in caves was not unusual, but these seemed different. Fortunately the workers called a local schoolmaster with an interest in such matters.

Heavy and curved limb bones were initially thought to indicate a deficiency in Neanderthals but are in fact a result of their powerful musculature.

described and illustrated the bones and discussed the situation in which they were found and their condition. The workmen had told them they were dug out from below two metres or so of deposits and had been accompanied by some stones and animal bones. This suggested that the bones were of considerable age. Furthermore Fuhlrott had noticed that the bones were invaded by mineral growths, a characteristic typical of other ancient animal remains which at this time were still considered by many to have been creatures associated with the Flood. In any event, such characteristics were generally taken as indicative of great antiquity.

The interpretation of the Feldhofer bones was Schaaffhausen's responsibility and although he realized that they were different from both ape bones and those of modern humans, these were still pre-Darwinian days. He scoured the literature looking for any previous comparable find but could find none. In the end all he could conclude was that the bones were the remains of some ancient Germanic race, even though they 'exceed all the rest in those peculiarities of conformation which lead to the conclusion of their belonging to a barbarous and savage race'.

However, 'sufficient grounds exist for the assumption that man coexisted with the animals founding the Diluvium [deposits of the Flood]; and many a barbarous race may, before all historical time, have disappeared, together with animals of the ancient world, whilst the races whose organization is improved have continued the genus'. Although hardly revolutionary ideas, by the standards of the day it was still highly inflammatory to voice such ideas of antiquity for mankind and progression.

Critical attention

The effect was to draw a lot of critical attention to the find and the discussion of what exactly the remains represented, and in the process Schaaffhausen was taken to task by some of the most eminent scientists in Germany.

One was Rudolph Virchow (1821-1902), a very great pathologist but also an enemy of evolution. As far as he was concerned, the bones, especially the leg bones – which show a distinct curvature – were just those of some unfortunate human who had suffered from a pathological condition such as a severe case

of prolonged and chronic Vitamin D deficiency, which produces rickets.

But it was Schaaffhausen's Bonn colleague August Mayer (1787-1865) who produced the most imaginative explanation. He reckoned the curvature of the leg bones were typical of a lifelong horseman (although rickets might also have been a contributory factor), and as regards the brow-ridge, as the subject's left arm had been broken and badly healed, the constant pain had produced such a persistent frown as to promote the growth of the furrowed bony ridge. To Mayer the victim was clearly a deserter from the Cossack cavalry which had encamped by the Rhine in January 1814 before continuing their pursuit and harassment of Napoleon's forces during the retreat back into France from their ill-fated Russian expedition. Wounded, the Cossack left the encampment and climbed up into the cave where he died of his injuries!

Like Buckland, Mayer and Virchow could only see what they wanted to see and were prepared to go to any length, however fanciful, to explain the obvious. It took courageous scientists like Darwin, Wallace and Huxley to break through the prevailing mindset.

Thomas Henry Huxley recognised the explanatory power of the Darwin-Wallace theory of evolution by natural selection and proved to be one of its most effective advocates.

Despite the academic 'putdowns' in Germany, the ensuing argument did at least bring international attention to the find. Thomas Henry Huxley (1825-95), who became the doughty defender of Darwinian evolution, discussed the Feldhofer bones in his 1863 collection of essays entitled *Evidence of Man's Place in Nature*. From a comparison of the skullcap with that of humans and apes, he concluded that it was the 'most pithecoid [apelike] of known human skulls' and formed 'the extreme term of a series leading gradually from it to the highest and best developed of human crania'. Furthermore he speculated that 'in still older strata do the bones of a fossilized Ape more anthropoid, or a man more pithecoid, than any known await the researches of some unborn palaeontologist' – how right he was.

The Feldhofer remains named

However, in the same year 1863 the Feldhofer remains were at last given a new scientific status and name – one they have retained to this day. In a highly speculative address to the annual meeting of the British Association for the Advancement of Science (published in 1864), William King (1809-86) an English palaeontologist and professor of geology in Queen's College, Galway, Ireland announced that there were sufficient distinctive features to the bones to warrant placing them in a new human-related species which he called *Homo neanderthalensis*. King had not even seen the specimens and was relying on the published description and figures. He thought the skull quite chimpanzee-like and nearly placed it in a separate genus.

So the first genuine extinct member of our genus was 'born' scientifically and members of the species have been known ever since as the 'Neanderthals'. The unjustified and unfortunate association of this name with brutishness, even idiocy, entered the language right from the start and has been attached to the name ever since. But as we shall see, there was a problem with such an association because when the scientific community finally accepted the concept of human evolution in the late 1870s and '80s it was thought that the Neanderthals might be the immediate ancestors of modern Europeans. The only way to be comfortable with this ancestry was to distance ourselves from the Neanderthals by time.

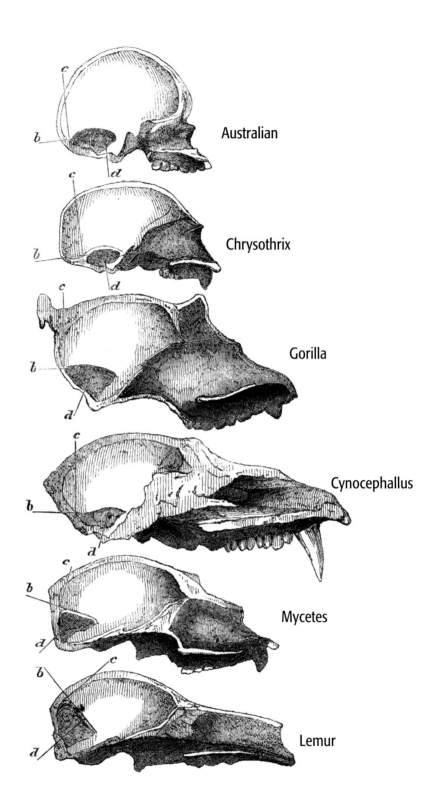

Australian

Chrysothrix

Gorilla

Cynocephallus

Mycetes

Lemur

Huxley's illustration of sections through a series of primate skulls shows how their shape was modified through evolution with a shortening of the snout and enlargement of the brain.

The association of the name Neanderthal with less desirable traits is unfortunate.

Two Neanderthal skulls from France showing the heavy browridge and low sloping forehead. The upper skull belonged to an almost toothless old man.

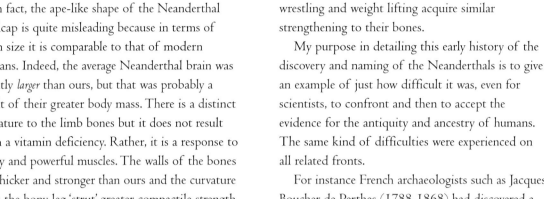

The apparent gulf between them and us also helped to make the arrival of modern humans all the more 'miraculous'.

In fact, the ape-like shape of the Neanderthal skullcap is quite misleading because in terms of brain size it is comparable to that of modern humans. Indeed, the average Neanderthal brain was slightly *larger* than ours, but that was probably a result of their greater body mass. There is a distinct curvature to the limb bones but it does not result from a vitamin deficiency. Rather, it is a response to heavy and powerful muscles. The walls of the bones are thicker and stronger than ours and the curvature gives the bony leg 'strut' greater compactile strength than a straight bone. The limbs were well adapted for heavy-duty carriage of a heavily muscled body over punishing terrain for long periods. Modern athletes actively engaged in power sports such as wrestling and weight lifting acquire similar strengthening to their bones.

My purpose in detailing this early history of the discovery and naming of the Neanderthals is to give an example of just how difficult it was, even for scientists, to confront and then to accept the evidence for the antiquity and ancestry of humans. The same kind of difficulties were experienced on all related fronts.

For instance French archaeologists such as Jacques Boucher de Perthes (1788-1868) had discovered a wealth of stone tools in the Somme River Valley area around St Acheul in Picardy in the 1830s and '40s. In 1847 he published a magnificent and lavishly illustrated volume entitled *Antiquités celtiques et anté diluviennes* in which he described the remarkable pear-shaped and bi-faced flint hand-axes (which have ever since been known as Acheulian) and the bones associated with them. Together they were good evidence of the coexistence of early tool-making human relatives with extinct (antediluvian) animals of the Flood. Again it confirmed what John Frere had found nearly 50 years before, but it still took a further 20 years and several more important finds for the truth to be accepted.

Images from the distant past

The discovery of the art works of our remote forbears is remarkably ancient, much more so than is generally realized. In fact, as long ago as the third century BC Chinese scholars first recorded the occurrence of images carved into rock surfaces (technically known as petroglyphs, literally meaning 'rock writing') in the open landscape. Many, but by no means all of the images are of stylized animals and are often carved on prominent cliff faces and some have been rediscovered in recent decades by working from the original location descriptions.

There are various European references, dating from the 15th century, to both rock and cave art, but it was from the 17th and 18th centuries, the period of European colonial expansion and exploration, that there are a number of records, some of which were published, from all around the New and Old World. Most, but not all of these accounts still refer to rock art out in the open. For instance, the Oxford Celtic scholar and Keeper of the Ashmolean Museum, Edward Lhwyd (1660-1708), sketched the rock art decoration of the passage rock into the Newgrange megalithic-chambered tomb in Ireland and described it in a letter to the Royal Society in 1699.

The work of savages – or Satan?

Observers were alternately puzzled and intrigued by the images and their possible meaning. Many dismissed them as the scribblings of savages or even the work of the devil, but some of the more enlightened scholars recognized their manifest artistry. However, they had no idea how old the work was and so the more aesthetically pleasing images were often seen as the work of Europeans rather than 'native savages' or aboriginal peoples from prehistoric times.

The scholars were so concerned about the sexual content of some of the humanistic images that they were not widely publicized and when illustrations were finally published they were heavily censored.

By the early decades of the 19th century the rock art of Scandinavia and Siberia was well known to scholars and increasingly reliable accounts and illustrations from regions as far away as Australia, Africa and the Americas demonstrated just how ubiquitous it was. However, in the Eurocentric world of archaeology, it was the European discovery of 'portable art' (meaning carved and decorated objects which were small enough to be carried), which

gradually drew more widespread attention to the work. The first European pieces of palaeolithic portable art known to have been discovered and illustrated in a publication were found in about 1833 in Veyrier Cave in Haute Savoie. These pieces comprised a carved harpoon-like antler and a perforated antler baton with an animalistic engraving on its surface. Again, it took several decades before the full significance and importance of all this evidence finally sank in.

The discovery by French palaeontologist Edouard Lartet (1801-71) and an English Banker and archaeologist, Henry Christy (1810-65), of carved bones and stones within cave floor deposits alongside the remains of extinct animals of the ice ages was finally accepted as indicative of their great antiquity. Perhaps the most famous of these early finds was made in 1864 in La Madeleine rock-shelter in the Dordogne, France. It was a piece of mammoth tusk engraved with a clear depiction of a mammoth.

Here was the image of a hirsute elephant with long curved tusks and a distinct 'topknot' elevation to the top of its head, all characteristics that distinguished it from today's elephants. It proved beyond doubt the coexistence of our ancestors with the extinct animals of the ice ages and showed that not only were they keen observers but some of them were also accomplished craftsmen. The downside of these discoveries was that it also initiated a 'gold rush' with indiscriminate excavation of cave deposits in search of such objects.

Mammoth engravings prove beyond all doubt our ancestors shared their world with animals that are now long extinct – and that early humans were keen observers of the world around them.

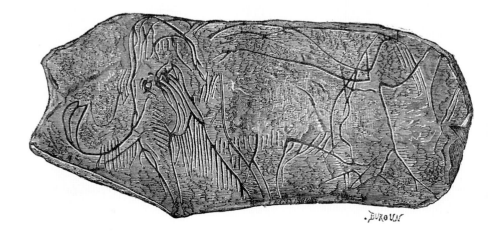

This 19th century copy of a Palaeolithic engraving shows an accurate depiction of the extinct mammoth with its small ears, long hair, long curved tusks and prominent forehead. The outline was engraved onto a piece of mammoth tusk found at La Madeleine in France.

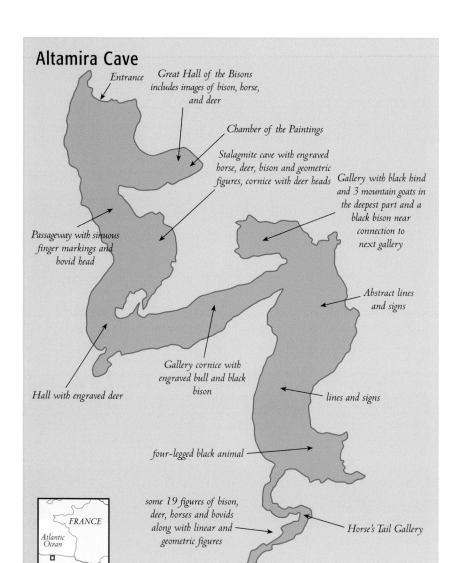

Altamira Cave

Entrance

Great Hall of the Bisons
includes images of bison, horse,
and deer

Chamber of the Paintings

Stalagmite cave with engraved
horse, deer, bison and geometric
figures, cornice with deer heads

Gallery with black hind
and 3 mountain goats in
the deepest part and a
black bison near
connection to
next gallery

Passageway with sinuous
finger markings and
bovid head

Abstract lines
and signs

Hall with engraved deer

Gallery cornice with
engraved bull and black
bison

lines and signs

four-legged black animal

some 19 figures of bison,
deer, horses and bovids
along with linear and
geometric figures

Horse's Tail Gallery

FRANCE

Atlantic
Ocean

SPAIN

0   25 metre

Altamira – The amazing quality of the famous polychromatic bison, the diversity of other animal images and abstract images make Altamira one of the most famous sites for cave paintings in the world. The 14,000 year old images are spread throughout the natural galleries and passages of this limestone cave, one of many used by Palaeolithic people in Northern Spain.

'Mira, Papa, bueyes'

Perhaps most important of all was a discovery made in Pyrenean Spain in November 1879. Maria, the young daughter of Don Marcelino Sanz de Sautuola, had accompanied her father, a local landowner and amateur archaeologist, into the cave of Altamira when he was carrying out some excavations of the floor deposits looking for pieces of portable art. According to her later account she was 'running about in the cavern and playing about here and there... 'Suddenly I made out forms and figures on the roof...mira, Papa, bueyes [look, Papa, oxen]' she exclaimed.

What she had seen was an amazing cluster of polychromatic bison that decorated the ceiling. Painted with ochre, outlined and shaded in charcoal black, they had remained there probably unnoticed

in the darkness of the cave for some 14,000 years since they were first painted.

But the ensuing story of the find was far from straightforward and caused so many problems for the proud Sanz de Sautuola that his daughter claimed that they brought about his early death. In 1880 he published a little illustrated pamphlet in which he claimed that the paintings were probably of a similar age to the portable art being found in cave deposits elsewhere. Despite some support from Spanish archaeologists and the king of Spain Alfonso XII, Sanz de Sautuola was effectively accused of forgery. French archaeologists such as Gabriel de Mortillet (1821-1898) and Emile Cartailhac (1845-1921) were particularly critical.

The images were so startling in their power and technical skill, and in such good condition that many experts could not believe that they were of any great age. Some experts argued that there should be smoke marks on the paintings because the only light available to the artists would have been from crude and smoky animal-fat lamps. The fact that there were no such marks fuelled suspicions that they were modern forgeries that had been painted with the aid of modern lamplight. We now know from experimental studies that some animal fats are not particularly smoky and so would not have left any sooty residues. And the artists could have put their lights on the floor which was originally much higher. It was subsequently dug down for the convenience of visitors who wanted to see the paintings.

More importantly, the paintings were generally dismissed as modern forgeries by the experts of the day largely because of the influence of de Mortillet who was virulently anti-clerical. He thought that the paintings, which he had not seen, were part of some anti-evolutionist Jesuitical plot to discredit the growing evidence that humankind had a prehistory. Not until 1895, and the discovery of engravings within La Mouthe cave in the Dordogne, where the cave's entrance had been blocked by sediment deposits and therefore effectively sealed, was Altamira accepted as genuine. But by then it was too late for Sanz de Sautuola who had died in 1888. In 1902 Cartailhac published a personal apologia entitled *Mea culpa d'un sceptique* in which he somewhat begrudgingly admitted that the Altamira paintings were genuinely as ancient as Sanz de Sautuola had claimed.

Questions of age

With rock and cave art finally accepted as the work of prehistoric humans, the archaeologists set about trying to make some sense out of its chronology and its function. Inevitably it was assumed that the more primitive the work, the more ancient it was, and the more sophisticated it was, the more recently it had been created, but it was all on a relative scale. Portable art at least had the chance of being related to the tools and bones with which it was sometimes found.

The same problem had occurred with the dating of stone tools and again the assumption was that the more primitive artefacts would be the older ones. What helped date them, however, was that stone tools were often found within layers of sediment along with animal bones. The identification of these allowed the development of a relative stratigraphy.

The French archaeologist Gabriel de Mortillet suggested that stone tools could be treated as fossils with particular tool types characterizing successive divisions of strata. Thus he divided the Palaeolithic into four, and then six subdivisions. From oldest to youngest the Lower Palaeolithic divisions were named as Chellean and Acheulian (characterised by bifacial hand-axes). The Mousterian has a greater diversity of axes and flakes for instance, and along with Neanderthal skeletal remains formed the Middle Palaeolithic. Then the Aurignacian (containing bone points and long thin blades), Solutrean (with fine 'laurel-leaf' points) and Magdalenian (with bone and antler tools, decorative items, bladelets and so on) successively form the Upper Palaeolithic and are associated with the skeletal remains of modern humans. Today the Châtelperronian is inserted between the Middle and

One of the spectacular polychrome bison that are painted onto the roof of a limestone cave at Altamira in northern Spain by late Palaeolithic people some 14,000 years ago.

Dating ancient art proved to be as challenging as dating remains and artefacts... in fact in many ways, it was more challenging.

These relatively simple stone tools are nearly two million years old and belong to the Oldowan culture of Africa.

Upper Palaeolithic and is associated with the interface between the Neanderthals and modern Cro-Magnon humans.

The sequence of stone tool technologies has generally been upheld by modern investigations and the advent of scientific dating techniques, although the relationship between the African and Eurasian sequences is more problematic. The original hope was that a similar chronological sequence could be established for the artwork. The baseline was taken as the arrival of modern humans in Europe. It was thought that the development of the art was a largely European affair despite the evidence that such work had a virtually worldwide distribution.

The fact that many of the works depicted the extinct animals of the ice ages also provided them with a younger chronological limit since the palaeontological evidence indicated that most of the bigger animals disappeared not long after the end of the last glacial and the beginning of Holocene times.

Famously, the French priest and cave art expert the Abbé Henri Breuil devised a four-fold scheme which he later reduced to a two-cycle one each developing from a crude to a more sophisticated style. There was an older 'Aurigno-Perigordian' cycle from which was derived the younger 'Solutreo-

Magdalenian' cycle. The former featured hand stencils, finger markings, outline animals without legs and animals in-filled with a simple colour wash eventually leading to bichrome figures. The latter cycle again begins with simple outline animals then black figures with infill and hatching and finally bichromatic images such as seen in Altamira. We now know that while this is generally so, it is not completely true.

A revolution in cave art chronology

A major problem with dating cave wall (parietal) art was the difficulty of relating the creation of the images to any particular stone tool technology as very rarely did any part of the art become incorporated in the sequence of cave deposits and associated tools and fossil bones. However, in the 1880s some evidence of relative age had come to light when some engraved cave walls, which were covered with Gravettian age deposits and associated artefacts, were found near Bordeaux. This proved that at least those particular artworks predated the Gravettian and it was this evidence that finally forced de Mortillet to accept the antiquity of the Altamira images.

Generally, though, there was no independent means of checking whether the proliferating ideas about its stylistic chronology were correct. However, with the advent of an advanced form of radiocarbon dating called accelerator mass spectroscopy (AMS, see p. 182) in the 1960s it became possible to date much smaller samples of carbon (100-500mg).

The natural pigments commonly used in cave paintings included red ochre (an iron oxide mineral called hematite), carbon black in the form of burnt bone and charcoal (and sometimes a manganese mineral) and occasionally white clay (kaolin). In a few instances enough carbon could be removed for dating without significant damage to the work. The results, however, were not always what was expected.

For instance, Chauvet cave in the Ardèche with its sophisticated charcoal drawings of woolly rhinos and big cats turned out to be over 30,000 years old – so much for the naïve idea that the primitive form would be oldest. Instead it confirmed what was already suspected from the early dating of some portable carvings, especially those from Aurignacian levels at Vogelherd and Hohlenstein-Stadel in Germany. The sophisticated carving of a human figure with a lion's head from the latter site has been dated at 31,750 years old.

Art history

So the big question was – how could such sophisticated art appear so early on without any apparent history of development? By the 1960s it was already evident that modern humans had arrived in the Middle East from Africa by around 90,000 years ago. So there was a significant period of time from which no 'art' had been recovered and there was the possibility that more primitive work could originate from this time if only we could find it. There are a number of possible reasons why examples are not known from this time.

Population numbers were very small as is the sample of skeletal remains of this age, thus the chances of finding any of their 'art' is slim. It was probably portable since the people lived in small groups of mobile hunters. And perhaps parietal art was only made when they had clearly defined territories with regularly visited caves.

In addition, there is then the question of the African hinterland and homeland of modern humans and the possibility that earlier and more primitive work was to be found in Africa. As we

A human figure with a lion's head carved from mammoth ivory is one of the oldest and most sophisticated of early Palaeolithic artworks known. It was found in the cave of Hohlenstein-Stadel in Germany in 1931 and is over 31,000 years old.

Carbon-dating some of the rock art provided results that were not always expected.

have seen, rock art has been known from Africa for well over 150 years, especially from the Sahara and southern Africa but mostly it is impossible to date. Furthermore many of the images include human figures which are largely absent from the early Eurasian art, suggesting that most of this African art does not date back more than a few thousand years.

There is certainly a preservational problem for portable art in much of Africa as organic materials such as wood are hardly ever preserved. Bone and ivory can be preserved if buried in suitable situations such as caves. However throughout much of Central and Northern Africa caves containing human-related remains are very rare. It is only in the limestone terrains of southern Africa that caves occur more commonly.

The story in the strata

Stratigraphy – the study of strata – is a fundamental technique of geology for establishing the relative age of strata and is based on the principle or 'law of superposition of strata', which, simply put, holds that in any sequence of **undisturbed** sedimentary rocks or layers, the oldest will be at the bottom, the youngest on top.

Accordingly, in any sequence of sediment layers, progressively younger layers cover the oldest and, since strata can be identified by their associated fossils, which change over time, it is possible to establish a relative chronological sequence with their characteristic fossils. This can be used to correlate between geographically separate sequences of strata. Here the development of major Stone Age industries is arranged against the stratigraphy found in Olduvai Gorge.

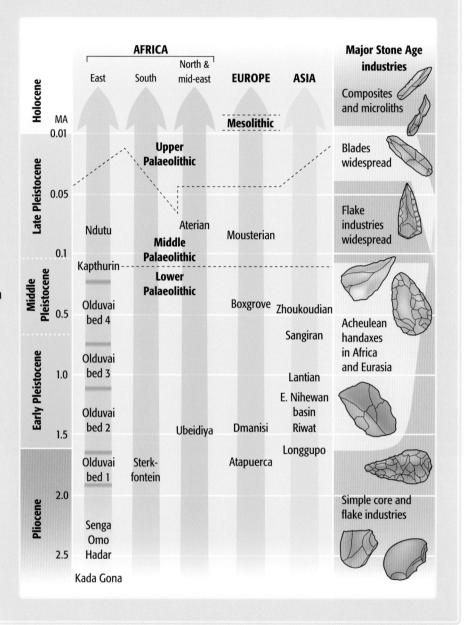

At 2.6 million years old these simple stone tools from Gona in Egypt are the oldest so far discovered but it is not clear who made them.

Blombos Cave

The idea that an apparent explosion of sophisticated art works in Eurasia might have a long and largely unpreserved African 'fuse' has received a considerable boost from recent 77,000-year-old discoveries in Blombos Cave in South Africa. The coastal cliff cave lies some 290km (180 miles) east of Cape Town and was excavated by South African archaeologist Christopher Henshilwood in 2001. Within a sequence of cave floor deposits, the excavation team found two small elongate pieces of red ochre about 5cm (2in) long with flattened sides on which a clear criss-cross X pattern of lines has been inscribed.

There is little doubt that the pattern is intentional and can only have been made by a human-related hand. There is a flat surface on one side of the crayon-like piece of ochre on which there a diagonal pattern has been scratched and its edges defined by further inscribed lines, all made by some sharp-pointed tool. Ochre is not particularly hard so a broken piece of bone, antler or pointed stone would have done the job. Moreover, the pattern has been repeated several times one on top of the other and it looks as if each layer of inscription has been partly obliterated by being rubbed against a hard surface and then re-inscribed.

The use of red ochre is intimately associated with modern humans, especially when used as a scattered powder on burials (such as Paviland in South Wales, see p. 73) as well as a pigment in cave art. It is highly likely that it was also used for body painting but we have no information about this latter usage in prehistoric times. The use of ochre at Blombos predates the oldest Eurasian art by around 40,000 years. Although its form is that of an abstract geometric pattern and may have been a simple tally or counting device it nevertheless is a concrete representation of a cognitive development for which we have no record from our more ancient relatives. The graphic symbols may well have been used as part of a transmission of information between individuals.

This is an important example of early symbolic behaviour even if it is not what we might normally associate with art. And, it is most likely to have been associated with the early modern human occupation of this site. Similar abstract forms accompany many of the animalistic images of Eurasian cave art.

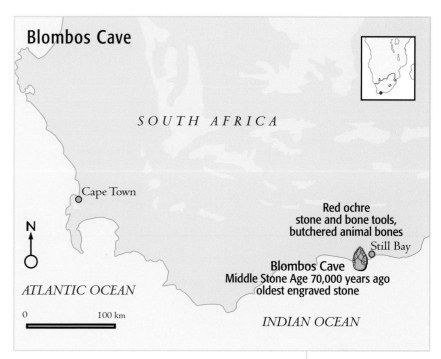

Blombos Cave

SOUTH AFRICA

Cape Town

N

ATLANTIC OCEAN

0   100 km

Red ochre
stone and bone tools,
butchered animal bones

Still Bay

Blombos Cave
Middle Stone Age 70,000 years ago
oldest engraved stone

INDIAN OCEAN

Although no human bones have been found at the site the ochre 'crayons' were associated with a cache of carefully worked bone points, stone tools and animal bones. The latter have been broken and show signs of defleshing and marrow removal. The bone points are characteristic modern human artefacts and the style of the stone tools is typical of the African Middle Stone Age, which succeeds the Acheulian and Oldowan in that continent from around 200,000 years ago until around 50,000 years ago. At the same time in Eurasia, the Neanderthal people were manufacturing the Mousterian type of tool.

There are a number of other coastal cave sites in southern Africa that were occupied by early modern humans but most of them were excavated many years ago without the kind of detailed attention to the sedimentary context that is now employed. Consequently other examples of early 'artwork' may have been overlooked but some caves still retain sediment, which will now be researched in great detail.

Blombos Cave – This tiny cave, now perched high above sealevel, was much closer to sealevel during a warm phase around 80,000 years ago in the last glacial. Stone and bone tools, butchered animal bones and pieces of red ochre, one of which is engraved with an abstract diagonal design, show that the cave was occupied by early modern humans.

Some caves could warrant a further examination as it is possible some examples of early art have been overlooked.

This 5cm long piece of red ochre from Blombos Cave in South Africa is engraved with the earliest known intentional marks made by humans some 77,000 years ago. The significance of the linear pattern is unknown.

A re-enactment of a Neanderthal scene. An actor with facial prosthetics shows how Neanderthals, who occupied a large tract of Europe and Western Asia between 200,000 and 28,000 years ago, used flint flakes to sharpen wooden spears.

PART II

Our extended genealogical search for our ancestry begins with our most familiar immediate ancestors and extends deeper and deeper into the remote and less familiar past until we reach the common ancestors we shared with the chimps over six million years ago. Curiously, for a species so obsessed with family and family history, very few people today have any idea who our evolutionary ancestors were.

CHAPTER FOUR

Our immediate pre-*sapiens* ancestors

 The scientific search for our immediate ancestors has been one of the most contentious issues in our discovery of our prehistory because it inevitably challenges received opinion. All cultures have developed some kind of story to explain our existence and most involve religion and a deity. Any threat to such ancient belief systems, especially when it questions fundamental issues of dogma, is bound to be resisted by those religious and cultural authorities who have a vested interest in maintaining the status quo.

And, when it was first realised that the Neanderthals might be our immediate ancestors the problems of accepting the scientific evidence did not diminish. But then the Neanderthals had been seriously misrepresented.

Neanderthals were successful hunters of game and worked together in small groups to ambush and kill animals as large as horse and deer with hand-held wooden spears.

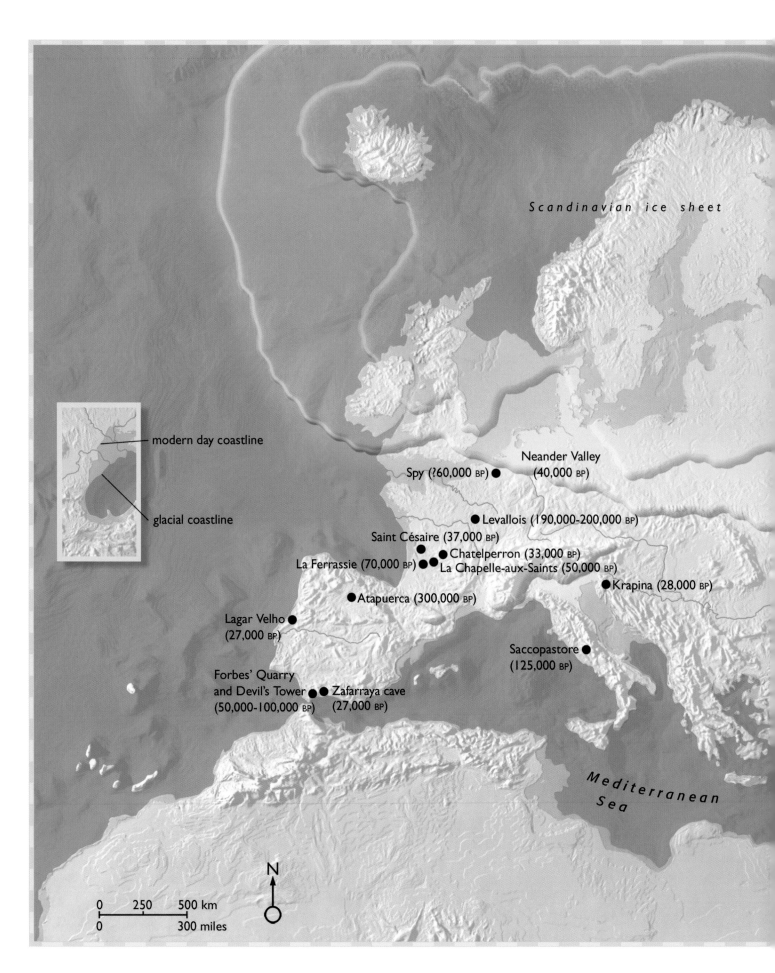

Scandinavian ice sheet

modern day coastline

glacial coastline

Neander Valley
(40,000 BP)

Spy (?60,000 BP) ●

● Levallois (190,000-200,000 BP)

Saint Césaire (37,000 BP)
● ● Chatelperron (33,000 BP)
La Ferrassie (70,000 BP) ● ● La Chapelle-aux-Saints (50,000 BP)

● Krapina (28,000 BP)

● Atapuerca (300,000 BP)

Lagar Velho ●
(27,000 BP)

Saccopastore ●
(125,000 BP)

Forbes' Quarry
and Devil's Tower ● ● Zafarraya cave
(50,000-100,000 BP)   (27,000 BP)

Mediterranean
Sea

N

0   250   500 km
0   300 miles

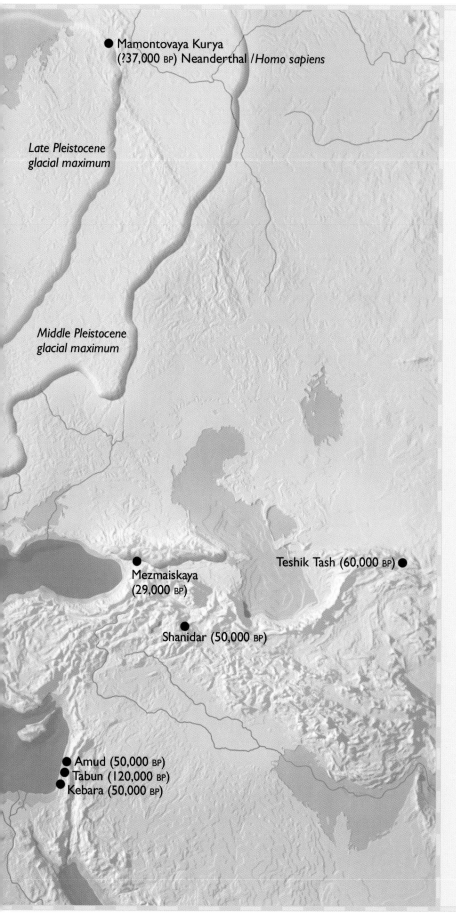

Mamontovaya Kurya
(?37,000 BP) Neanderthal /*Homo sapiens*

*Late Pleistocene
glacial maximum*

*Middle Pleistocene
glacial maximum*

Teshik Tash (60,000 BP)

Mezmaiskaya
(29,000 BP)

Shanidar (50,000 BP)

Amud (50,000 BP)
Tabun (120,000 BP)
Kebara (50,000 BP)

Neanderthal territory

The first Neanderthal remains were found in Gibraltar in the early 19th century but were not recognised as such until much later. It was the German find in the Neander Valley near Dusseldorf in the 1830s that gave its name to the species *Homo neanderthalensis*. This was the first extinct human species to be described and it was many years before it was accepted as such. At the time it was not possible to chronologically date archaeological material but it was clear that they lived alongside many of the animals of the Ice Ages and were replaced by the first modern humans so there was a question over whether the Neanderthals could have been the immediate ancestors of modern Europeans?

Today, Neanderthal remains have been found over a large area of southern Eurasia from Germany south to Spain and east to Israel and northern Iran. However, it is quite likely that they extended beyond this region but we have yet to find their remains. There is one site known from much further north within Arctic Russia but it is not clear whether it was Neanderthal or early modern human.

Although they lived through the latter part of the Ice Ages and may have been quite well adapted to the cold, there is now evidence that they moved south when the climate deteriorated during cold glacials and only extended back north when conditions improved. The Neanderthals survived in the region for over 250,000 years before dying out around 28,000 years ago. In their latter millennia their territory was encroached upon by incoming modern humans. It is not known what interaction, if any, there was between the two human species but the outcome is known, the Neanderthals died out without any surviving genetic 'issue' within the gene pool of modern Europeans.

Neanderthals are us?

The Neanderthals were the first of our extinct relatives to be recognized as a separate and ancient human species — *Homo neanderthalensis* in the latter half of the 19th century. As we have seen in the previous chapter, acceptance of their distinctiveness took some decades as questions were raised about their relationship to modern humans of European origin. Was there perhaps a genetic 'shadow' of the Neanderthals in the European inheritance? Only in recent times has this possibility finally been discounted.

Right from the start the Neanderthals were characterized as being generally 'nasty, brutish and short' savages of low intelligence verging on idiocy. This character assassination remains with us in the common usage of the name 'neanderthal' in a derogatory sense. In sociological terms, they seem to have become the archaeological equivalent of the 'idiot community' that every culture uses as a butt of their jokes.

Neanderworld

The truth is very different... We now know that the Neanderthals were a remarkably successful and very interesting group of our ancient relatives. Their homelands stretched from Britain east to Teshik Tash in the foothills of the Caucasus Mountains, south to Iraq and west through Palestine to Gibraltar. This huge area is just a minimum estimate based on known Neanderthal-related sites and was occupied off and on as climate change allowed for over 250,000 years, from around 300,000 to 28,000 years ago when the Neanderthals finally became extinct.

Anatomically our picture of the Neanderthals was biased by the selection and description of the arthritic skeleton of an 'old' Neanderthal man (actually in his mid 40s) as typical of the species. The skeletal remains were found by three priests at La Chapelle-aux-Saints in the Corrèze district of southwest France in 1908. Instead of being sent for study to the Ecole d'Anthropologie, they were sent to the less experienced anatomist Marcellin Boule. Perhaps this is not surprising since the Ecole had a reputation for anticlericalism inherited from its founder by Gabriel de Mortillet. Unfortunately, although Boule was aware from the skeleton that the 'old man' suffered from osteoarthritis, he claimed that Neanderthals could not walk fully upright and his reconstruction emphasized a round-shouldered and hunched look with an ape-like face and hairy

Side by side, the differences between Neanderthal (left, La Ferrassie, France) and modern human (right, Cro-Magnon, France) skulls are readily apparent. The face is larger with a prominent browridge and smaller forehead in the Neanderthal but brain size was similar.

body for which there was no evidence at all. There could not have been a greater contrast than with the emerging picture of the Cro-Magnon modern humans. It was not until the 1950s that the 'old man's' skeleton was fully re-examined and the extent of the false image revealed.

But now we have a much better idea of how these very interesting close relatives looked. Altogether the remains of some 500 Neanderthals have been found over the last 150 years, although most are very fragmentary and there are only a handful of near complete skeletons. Nevertheless, the interdependence of skeletal elements (the bones) means that calculations can be made for body size and mass from such isolated material. As we have seen, their skeletons are robustly built with thick-walled and slightly curved limb bones that developed this way in response to a very strenuous lifestyle as hunters of medium to large game animals.

Proportionally, Neanderthal limb bones are similar to those of modern humans who live in high latitudes, such as the Inuit, Lapps and Siberian tribes-people. Their forearms and shins are relatively shorter than is the global norm and typically the trunk of the body is thickset, with a barrel-shaped chest – and the same is true of the Neanderthals. Today these proportions are seen as an adaptation to living in cold climates and it has been argued that the same applied to the Neanderthals. The effect is to reduce the ratio of the surface (skin) area to body volume, which is a heat-saving adaptation; the shortening of the extremities reduces the risk of frostbite.

These adaptations have been taken to indicate that the Neanderthals could survive in the near arctic conditions of the ice ages, but as we shall see this is not really true. Certainly, they have a distinctive body form by comparison with the Cro-Magnon modern humans whom they encountered in western Eurasia during the latter part of their existence. However, in recent years the idea that the Neanderthals were cold-adapted has been challenged as their particular anatomy can alternatively be seen as an adaptation to their very rigorous lifestyle as active hunters, armed only with hand-held wooden spears, pursuing dangerous game on foot.

Accordingly, they needed to be tough and muscular, heavy boned with big lungs – like American footballers or rugby players. Perhaps the adaptations served both lifestyle and harsh climatic conditions.

From the limited available information male Neanderthals grew to a maximum of 179 cm (5ft 10in) and the smallest females were 155 cm (5ft 1in) This might make them seem somewhat 'vertically challenged' by comparison with men in the West today.

However the Neanderthal body mass of both males and females averaged 76kg (167 lb) – some 30% higher than our modern average – and none of the average Neanderthal's weight was 'flab', whereas a modern human of 179cm and 76kg would either be distinctly tubby or a well-muscled weightlifter or wrestler.

The remarkable brain size of the Neanderthals was generally ignored in early discussions. Even the famous 'old man' of La Chapelle-aux-Saints had a brain size of just over 1600cc and and Neanderthal brain size ranged from around 1300cc for females

Comparison of skeletons show how the Neanderthal build (left) differed from that of modern humans (right). The Neanderthals were shorter and stockier and powerfully built with relatively shorter lower limb bones. It was thought that the Neanderthals were slouch shouldered and bow-legged (far left) but this was based on a misinterpretation of the skeleton of an elderly individual with arthritis.

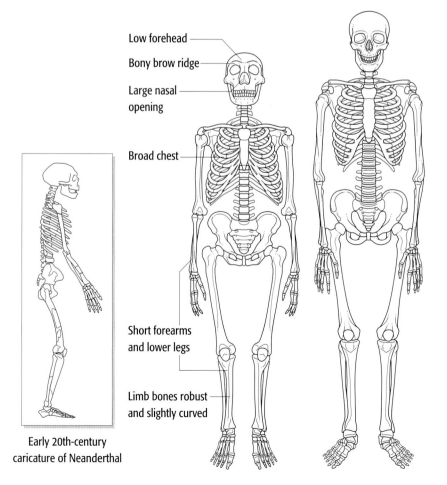

Low forehead

Bony brow ridge

Large nasal opening

Broad chest

Short forearms and lower legs

Limb bones robust and slightly curved

Early 20th-century caricature of Neanderthal

Here, a modern human actor is wearing a Neanderthal-style facial prosthetic which makes his brow prominent and greatly enlarges his nose. It is assumed that Neanderthals probably lost the original African skin pigmentation and acquired a paler Mediterranean type colouration over the many thousands of years they occupied northern European latitudes.

and 1600cc for males rising to 1740cc in the Amud man. These measures are slightly greater than those seen in modern humans (1200-1700cc, with an average of 1350cc) and show similar differences between the sexes as seen in modern humans. However, as we have seen, brain size relates to body mass and if we make a correction for the higher mass of the Neanderthals their relative brain size drops slightly below ours – but not much.

So, if they had such large brains how come they have been commonly regarded as being unintelligent – to put it politely? To try and answer this we need to look at some other Neanderthal characteristics and our understanding of their behaviour.

The Neanderthal skull and face have several distinctive features.

As we have seen, the shape of the skull roof was evident right from the first finds at Neanderthal in Germany. There is a strong bony brow-ridge that arches over the eye sockets. It was thought that this primitive feature, seen in so many of our ancient relatives and in the higher apes, acts as a strengthening to the skull, but mechanical tests have shown that it makes little or no such difference. There are various other claims for its function, such as a sun-visor, but none are particularly convincing. Interestingly it did not develop as a structure until adolescence and may even have been a secondary sexual feature.

Neanderthal 'big faces'

The Neanderthal face is relatively larger than that of modern humans and this is emphasized by the lack of a vertical forehead. The opening for the nose is also large and this central part of the face is pulled forward. It seems that the Neanderthal nose was significantly enlarged and this may have been a kind of 'air conditioning' cold adaptation since the big nasal passages would have warmed the air before it entered the lungs and this would have been advantageous during strenuous activity in cold weather. Alternatively, it could have simply been a means of improving air intake to large lungs.

The chin area is recessive, unlike that of modern humans, but this does not mean that they had a weak chin, which we still tend to associate with 'feebleness'.

The chin junction of the two bones of the lower jaw is not very strong and has to be reinforced by additional bone growth. In the Neanderthals and all our other ancient relatives that extra bone lies *inside* the chin while in modern humans it is external and produces our distinctive forward jutting chin.

Teeth and leatherwork

Neanderthal teeth are relatively large and often have distinctive wear patterns; this is particularly evident in the case of the front teeth, the incisors. The wear facets are not those produced by normal occlusion of the upper and lower jaw, but they are similar to those observed in Inuits – who process animal skins using their front teeth as a vice or clamp thus leaving

A reconstruction shows a Neanderthal woman holding an animal skin in her teeth, aided by a child. This leaves her hands free for scraping fat from the skin as part of its preparation as clothing. Wear facets and cut marks on Neanderthal teeth show that organic materials such as skins were frequently 'clamped' in the teeth and perhaps 'worked' to make them supple. The sloping cut marks show that most Neanderthals were right-handed like most modern humans.

one hand free to scrape the fat from the skin. The skins are then made supple by working them with the lower jaw being moved back and forth. As a result there are wear surfaces that slope down and out on the lower incisors and up and out on the upper incisors. In addition some upper incisors have slanting scratches made by the stone blades used to scrape the skins with and these show that most Neanderthals were right handed.

So, there is little doubt that they were habitually processing skins presumably for clothing, bedding or as part of temporary shelters. However, the absence of any bone needles among their artifacts shows that if skins were used for clothing they were not sewn together in any sophisticated way. Such advanced clothing technology appears to be an innovation of modern humans.

Bodyworks

The Neanderthal upper body or torso features thick and tough rib bones that form a typically barrel-shaped chest that enclosed large lungs. The bones of the torso often show signs of damage, which has been compared with that seen in modern rodeo-riders and is a result of being buffeted and kicked by animals such as horses or cattle. Neanderthal hand

bones are large and often curved as a result of powerful muscles. Right arm bones tend to be slightly larger than left arms showing that there was a habitual imbalance in the sort of work the right arm was doing.

This fits well with other evidence that indicates they typically used heavy 2m (6ft 6in) long wooden spears for hunting (see box on spears). Sometimes they were tipped with stone points (see box on Neanderthal glue) but in the absence of suitable rock material the points were fire hardened.

Recently though, some fragments of birch resin associated with Mousterian tools typical of the Neanderthals have been found in Germany. Most resins from pine trees is not suitable for firmly gluing sinew of leather bindings around stone points because it does not harden properly. However, birch resin does harden — but it has to be carefully processed and this involves heating to just the right temperature. Too little and it will not set, too much and it will burn.

This evidence of Neanderthal use of birch resin suggests that here they had developed a sophisticated technology normally associated with modern humans and not seen in any older human-related species.

The prominent brow-ridge might have been a secondary sexual feature, as it did not develop until adolescence.

The Neanderthal spear was a long and heavy wooden shaft that sometimes was simply pointed but at other times were tipped with sharp stone points. The spear was too heavy to throw and was probably used like a bayonet for thrusting and stabbing a prey animal. But first the Neanderthal hunter had to get close enough.

Spears – one of humankind's earliest weapons

Organic materials such as wood do not easily survive as archaeological artefacts except under certain special circumstances such as in oxygen-poor mud and water. Consequently, the record of the prehistoric use of wood is very limited, although it must have been one of the earliest natural materials to be employed for a number of peaceful purposes, such as digging. Of course, it was also used for more aggressive purposes including hunting, and defence and attack.

In the same manner, some of today's chimpanzees use wood for such basic purposes and it is likely that our common ancestor also did so over 5 million years ago. However, the earliest record of the use of wood as a weapon dates back a mere 400,000 years or so and consists of eight remarkably well preserved wooden spears found along with the butchered remains of some 20 horses. The find was made in Schöningen, eastern Germany.

There is also indirect evidence from Boxgrove in the south of England for the use of wooden spears. This evidence dates back even further, to some 500,000 years. It consists of circular holes punched through the shoulder blade of an extinct species of rhinoceros that was hunted, killed and butchered. It is most likely that wooden spears made the holes. But at Schöningen being covered by waterlogged marsh deposits has preserved the spears themselves.

They are between 1.8 and 2.3 m (6–7.5 ft) in length and are made of spruce with simple pointed ends rather than being tipped with stone points. Stone tools were also found at the site but they were used for butchering the animal carcasses. Despite not being tipped with heads of stone, however, the spears were still formidable in skilled hands as the tips of wooden spears can be hardened by the careful use of fire.

The world's first superglue

Working out exactly how our ancient relatives attached stone points to wooden shafts to make spears, and later arrows, was something of a problem for many years. The very rare preservation of various types of bindings (such as those found with Ötzi the 'Iceman', see p. 54) has since provided some answers.

A time of danger – the kill

Neanderthal spears were too heavy to be thrown and were instead used as thrusting bayonet-like weapons in hunting. That meant the Neanderthal hunters had to get sufficiently close to kill their prey with these spears. We know that they hunted and killed medium-sized game such as horse, wild cattle and deer as well as big game such as mammoths. Such animals are not easy to kill with a single spear thrust and it would require the close cooperation of a few hunters who had to get very close to the prey – and that posed a hazard...

Frightened and wounded animals of this size can be highly dangerous as a single kick could easily break a hunter's leg or arm. The evidence of torso damage in Neanderthal skeletons supports this interpretation of their hunting style.

Of course, simply getting close enough to the target posed its own challenges... Their prey animals were mostly fleet of foot, very wary, and equipped with acute senses of smell, hearing and sight, so that Neanderthals would have had to be highly proficient at stalking the game. Their hunting strategy may well have included driving the animals into natural traps where they would be easier to kill. However, this requires quite a number of hunters working as a team, and all the evidence suggests that Neanderthal bands numbered not many more than a dozen individuals including women, children and the aged, so the number of active hunters may not have been more than a handful. Our ideas about their group size come from analysis of the sort of area covered by known Neanderthal camp, butchering and tool making sites.

Neanderthal lifestyles

Since much of the game that the Neanderthals hunted was highly mobile and some of it migratory, the hunting teams would have had to travel significant distances from the home base. It is likely therefore that the teams consisted primarily of fit males as inexperienced young would have been a liability and the more experienced elders would have been too slow. The difference in build between the males and females was slightly greater than in modern humans and again it is unlikely that the females took part in long-range or potentially dangerous hunts.

However, as mentioned earlier, in their turn the females were powerfully built and their skeletal remains show some upper body damage as well. This suggests that perhaps they too took part in some

Some evidence suggests that Neanderthal females might also have taken part in hunting.

There is good isotopic evidence from Neanderthal teeth and bones that they were serious flesh-eaters and therefore heavily dependent upon hunting game animals, especially in the winter.

dangerous hunts, presumably when they were young and not heavily pregnant. It is also safe to assume that when their children were very young and still breast feeding (see below) the mothers could not have accompanied the hunters but instead probably caught small game locally and gathered plant-based food in season – fruits, tubers and so on. In fact, it was probably the females who held detailed knowledge of such foods, where to find them and how to prepare them.

In addition to the above-mentioned pursuits, they were very likely employed with other tasks such as preparing skins resulting from earlier hunting successes.

We know from isotope studies that Neanderthals were serious meat eaters – on a par with big cats and wolves. The advantage of such a diet is that it is high in protein and provides energy quickly, especially if the meat is 'preprocessed' to some extent by being cooked. The presence of hearths and burned bones at Neanderthal sites shows that they regularly used fire to cook meat as well as provide warmth and protection against predators such as big cats and wolves – which were not only dangerous but also their competitors for some of the game.

A major problem was that kills were normally made some distance from the home base. Consequently, the hunters would have to butcher their prey on the spot and carry what they could back home. Butchering requires sharp stone tools and so a supply of these would have had to be

Although the stone tools that the Neanderthals manufactured were not particularly sophisticated, they did include handaxes and a variety of blades and scrapers for skinning and defleshing prey animals and scraping fat from their skins. Some of the more complex tools could be retouched but others were easily made and just discarded when they became blunt. Flint flakes can be very sharp but they are brittle and soon lose their edge.

carried on the hunt or picked up from known caches within their hunting territory.

Moreover, the tools themselves frequently needed replenishing or reworking to keep them sharp and that meant there was a constant need for more supplies of core stones from which new tools could be made. Depending upon the geology of the landscape in which they were operating, the supply of appropriate rock could vary enormously. Careful studies, especially in France, show that the Neanderthals would commonly travel up to 10km (6 miles) or more, and occasionally much further, to visit good rock sources.

In order to survive, a band of one to two dozen Neanderthals would require a kill of an animal the size of a horse every two weeks or so – somewhere around 20-30 beasts a year. To maintain such a supply each band would have had to range over an area of some 100km^2 (40 miles^2) so they would be thinly dispersed with relatively low population numbers, perhaps similar to the other large major mobile predators of the time – the big cats. However, the problem with a pure meat protein diet is the lack of fat, especially in the lean meat derived from active game.

So, to supplement this Neanderthals regularly sought out bone marrow from animal long bones, which they smashed open and consumed raw on the spot. Another dietary supplement was another easily digestible source of protein – brains and not just animal brains...

There is evidence of cut marks inside Neanderthal skulls, which show that the brains were removed and the most obvious reason would be as a food source although we cannot be sure. However, other Neanderthal limb bones have been broken open for their marrow, so it certainly does seem that they indulged in cannibalism. Furthermore, there is evidence that they practised infanticide (see below). Like some modern peoples living in marginal existences within stressful environments, they needed to constrain their group size and the number of mouths that could reasonably be fed. The killing of the newborn would be one solution.

The question of what happened to old Neanderthals is an interesting one.

Were they abandoned to die once they were past their 'sell by' date or even killed and consumed?

Archaeological evidence from tool debitage and animal bones with cut marks found at Neanderthal cave sites shows that kills were butchered and defleshed outside the cave entrance by small groups of a few individuals. Scatterings of used stone blades and flakes have been found surrounding the spaces occupied by each individual.

There is some evidence of care for the sick and wounded and the elderly may also have been tolerated so long as they were not too much of a burden on the group. They could have been useful in looking after children and foraging for firewood, catching small game and other 'household chores'. They may also have been useful repositiories of information provided their speech was sophisticated enough to transmit the knowledge but we do not really know.

If this model is correct, there are some interesting further implications for Neanderthal lifestyle and behaviour. Did Neanderthals have speech? Did the separate bands ever meet up? How did they prevent inbreeding?

The unique chance preservation of a hyoid bone from a Neanderthal throat shows very little difference to the modern human hyoid bone whose form is associated with the production of articulated speech. However, the overall structure of the Neanderthal throat and mouth region is slightly different from ours with the larynx being positioned slightly higher up than ours. This position is similar to that of a chimpanzee and modern human baby

The pitch and range of sound produced by primates depends largely upon the structure of the vocal 'instrument' – the throat and especially the position of the hyoid and larynx. Chimps and human babies have high pitched voices whilst Neanderthals have a similar vocal equipment to us and were probably capable of speech.

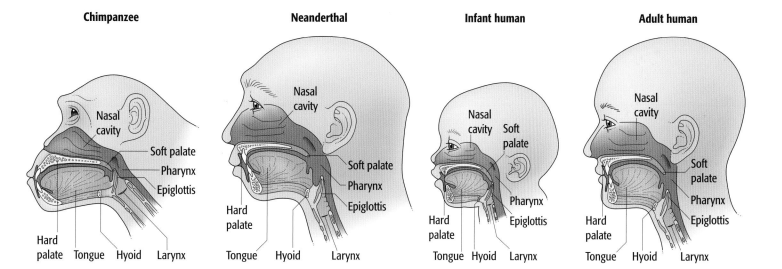

who cannot articulate clearly until the larynx has descended deeper into the throat. It seems likely that the Neanderthals could articulate a greater range of sounds than chimpanzees and they may have had similarly quite high-pitched (again like a human infant and child) voices.

Experiments have shown that although chimpanzees cannot speak they have all the cognitive abilities necessary to master basic language with vocabularies extending to over a hundred words. There is no doubt that Neanderthal cognition was far greater than that of chimpanzees and so it is likely that they had basic speech but not complex grammatical language. The possession of the latter is probably one of the key adaptations, which gave modern humans the edge over all their predecessors.

But there are other means of communication apart from speech: body language, facial expression, hand gestures and other vocalization such as calls and whistles can all be used to convey information. For active hunters working in small groups hand gestures, calls and whistles can be very effective means of communication. Chimpanzee bands hunt very successfully without complex language and rely on relatively simple calls and learned behaviour plus practice. So the argument goes that Neanderthals and our more ancient and smaller-brained relatives could also have managed without syntactical language.

However, there is another line of argument which suggests that the success of the Neanderthals – and remember they did survive as a species for at least 250,000 years – was in part based on simple speech. The great advantage of speech is that it can speed up communication and learning which otherwise has to be a prolonged process of observation, mimicry and practice through play. Ape and human babies and infants are all acute observers and mimics and they all love to play.

It takes a chimpanzee some five years or more to learn complex tool use such as cracking hard-shelled nuts with a hammer stone and anvil because it all has to be done by observation of the mother's behaviour and technique plus personal practice. There is fossil evidence from the skeletal remains of Neanderthal children, especially detailed studies of their teeth, showing that like chimpanzees they developed rapidly during the first five to seven years — more

rapidly than our modern human children in fact. The same is true of chimpanzee babies of today – they are more advanced and proficient at many skills than human babies until about the age of three when human development overtakes them. Neanderthal 'children' also developed rapidly and for longer than chimpanzee 'children'. Modern analysis has shown that development of the Neanderthal skull and face was already well established in two-year-olds and is measurably different from that of modern humans even though features such as the bony brow ridge are not apparent.

It is likely that there was an important advantage to the Neanderthals in having 'children' who did develop so rapidly to begin with (called precocial development) because it made them more independent of their mothers as quickly as possible. The sooner they could contribute to the 'welfare' of the group the better so there may well have been a premium on learning and that would have been facilitated by speech. Simple speech, which allows directive instruction, can help short circuit otherwise prolonged learning purely by example. It allows for cultural transmission of information by which means a body of knowledge can be built and passed from generation to generation. For hunters pursuing migratory game through diverse landscapes and environments the build-up and transmission of any predictive information about the likely behaviour of the game and the weather can provide great advantages and increased survivability.

For instance, the development of hunting and domestic 'hardware' from stone tool manufacture to making fire can be enhanced if successful new techniques can be easily passed on. Neanderthal tool manufacture was quite sophisticated and went far beyond the simple modification of one piece of stone by another. They had a diversity of tools from axe heads and points to blades, the manufacture of which required a prior mental image of what was required and how it was to be achieved through a complex process of manufacture which involved a series of different steps.

Importantly, the awareness and knowledge that at least one intermediary step is required indicate a significant development in cognition and mental processing. For example, initially a suitable block of appropriate stone has to be selected. Then a core of

Like chimpanzee babies, Neanderthal 'children' developed rapidly, but they continued to develop for longer than young chimpanzees.

the right shape has to be prepared from it. The next step involves striking pieces from that core; finally touching up will provide the final tool shape.

Passing on knowledge of such complex manufacturing processes requires a lot of time but that can be speeded up with some verbal instruction.

Ganging up

We do know that Neanderthal bands must have met up from time to time for the exchange of 'goods' and perhaps 'personnel'. A few French Neanderthal sites contain occasional seashells, which are perforated as if for use as a pendant or necklace.

These are not the only items of Neanderthal personal adornment (see box) but their significance in this context is to do with communication between

groups. The shells have been found at sites hundreds of kilometres from the sea and the most likely explanation for their presence so far inland is that they were traded between groups, perhaps when they met at favoured stone 'quarry' sites.

Some unique and readily identifiable rock types have been found to be distributed over areas of several hundred square kilometres, suggesting that they too were probably traded. Although some groups seem to have been prepared to travel quite significant distances to obtain favoured rock material. The exchange of 'personnel' in the form of daughters may well have been another reason for Neanderthal bands to occasionally 'gang up'.

Evidence of an appreciation of ornamentation?

Neanderthal ornaments – The French site of Grotte du Renne at Arcy-sur-Cure is famous for its strange mixture of stone tools, items of personal adornment and what looks like the foundations of a structure that may have been a shelter or hut, all dated at around 35,000 years ago.

Although normally such an assemblage would be associated with Upper Palaeolithic modern humans, here they are thought to be the work of Neanderthals. Grooved and perforated animal teeth look as if they were used as pendants and were found with tools known as 'backed knives'. These are typical of the problematic Chatelperronian industry that is arguably associated with a late Neanderthal development. Some experts argue that they were only 'aping' the new 'fashions' worn by modern humans who had arrived in Western Europe by this time.

Neanderthal sex

One interesting question about Neanderthal 'family' or group life is how, with such small numbers in the group, they managed to avoid the deleterious genetic effects of inbreeding through incest. A useful model is provided by some of the higher apes like the mountain gorillas, which similarly live today in small mobile groups and show marked differences in size and strength between the males and females (technically known as sexual dimorphism).

Females of reproductive age within such a group effectively form a harem and are jealously guarded by the dominant (silverback) male, which tries to keep any other male from impregnating them. Female gorillas are only infrequently fertile and so the task is somewhat simplified for silverback gorillas.

By contrast, if Neanderthals were like modern humans, the females would have been more or less continuously fertile except when they were breast-feeding – when ovulation is suppressed. In chimpanzees the onset of fertility is suppressed until the young females leave the family group and attach themselves to a less genetically related one, thus avoiding incestuous reproduction. It is likely that similar exchange took place between Neanderthal groups but we do not know the details – whether such an exchange was socially mediated or not. Apart from the genetic benefits, the transfer could bring cultural benefits in that the young female would bring a learned 'knowledge base' which she would then transfer to her own offspring.

Although the use of personal ornaments is mostly a characteristic of our species, a number of simple ornaments, such as pierced seashells have been found at inland Neanderthal sites. The shells must have been traded and passed from one Neanderthal group to another. Some experts argue that the Neanderthals were copying this behaviour from modern humans when the two species briefly overlapped in time before the Neanderthals became extinct.

Neanderthal origins

As a species the Neanderthals are a particularly interesting because of their recency, their restricted geographical range in western Eurasia and their temporal overlap with modern humans. The question of their origins is not easily settled and requires the examination of a number of other finds which have been made over Europe (see Box) and further afield in the last hundred years or so. Most important of these is *Homo erectus*.

Pit of bones

In addition to the remarkable finds of Gran Dolina, the limestone caves of the Sierra de Atapuerca in northern Spain have preserved another extraordinary find known as the Pit of the Bones (Sima de los Huesos). Here over 2000 bones belonging to some 32 individuals have been recovered since 1992 and represent the biggest hoard of fossilised human bones ever found. The bones include the nearly complete skulls of children and adults and a jumble of other body parts that mostly belong to teenagers.

Dated to around 300,000 years old, they show a combination of ancestral and more modern derived features which seems to place them somewhere between *Homo heidelbergensis* and *Homo neanderthalensis* thus possibly representing an evolutionary transition between the two species.

Why so many human bones fill this ancient pit that lies deep within a cave system is still a mystery. The cave was not occupied by these people and many of the bones show cut marks showing that they have been butchered. The remains may be the result of dismemberment as part of some unknown mortuary practice before the bodies were dragged deep underground and thrown into the pit.

Haeckel was not afraid of speculation, claiming that humanity's homeland had sunk beneath the waves.

Ernst Haeckel was a charismatic German evolutionist who supported Darwin's ideas in general but disagreed with him over the location of primate origins. Haeckel thought that the gibbon was the most primitive primate and that the 'missing link' between apes and humans would be found in Southeast Asia rather than Africa as Darwin argued.

Java Man – *Homo erectus*

The discovery of Java Man is perhaps one of the most extraordinary stories in our saga and has its origins back in the latter half of the 19th century.

One of the most powerful and charismatic proponents of Darwinian evolution in Europe was a German biologist and artist by the name of Ernst Haeckel (1834-1919). As a young man he had made an uninvited 'pilgrimage' from Germany to see his hero Darwin. As a very private person Darwin did not take too kindly to strangers turning up at his country home of Down House in Kent, but eventually relented when he heard that the young man had come all the way from Germany. However, Haeckel also had many of his own ideas especially about the origin of mankind.

In the 1860s he made a study of an old collection of human skulls from around the world amassed by Johann Blumenbach (1752-1840), and following the biological fashion of the day, tried to arrange them in order from the most 'advanced' to the most 'primitive'. The criteria were by modern standards arbitrary and thoroughly misleading with what were considered to be more simian features being taken as the most primitive and the most advanced being based on – guess what – a Caucasian ideal as perfected in classical sculpture.

Not surprisingly white Europeans came out top and native Australasians and Papua-New Guineans at the bottom.

Haeckel also considered that the gibbon of the southeast Asian islands was more closely related to modern humans than the African apes, which Darwin preferred. Haeckel constructed a map of ethnic origins according to his hierarchical scheme and drew in flow lines connecting the succession.

From this mapped distribution he concluded that human ancestry was derived from some location well to the east of Africa. However, it could not be India because most of the peoples of that continent 'scored' quite highly in his scheme. But Haeckel was not afraid of speculation and so he claimed that the homeland for humanity had been an area of land somewhere in the Indian Ocean. He called it Lemuria or 'Paradise' a landmass that had since sunk beneath the waves. Haeckel even gave a speculative name to the 'missing link' between the gibbon and humans – *Pithecanthropus alalus* – meaning the 'ape-man without speech'. He, like so many of his contemporaries, thought that humanness was based on the acquisition of upright walking, speech and increased brain size and also located this ancestor in time – somewhere in the late Tertiary (Cenozoic), either in the Pliocene or Miocene epochs.

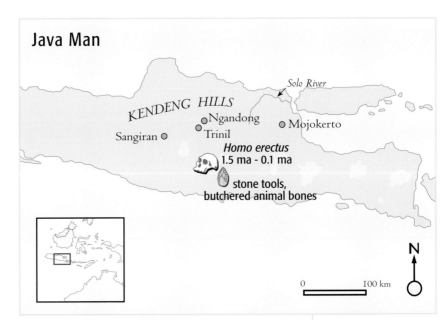

Java Man

In 1881 Eugène Dubois (1858-1940), a young Dutchman from the Catholic region of South Limburg, was appointed as assistant anatomist in the university of Amsterdam. A keen naturalist and budding scientist since his childhood, Dubois was soon publishing some important work on the anatomy of the larynx and had a promising career ahead of him. But on October 29th, 1887 he set sail for the Dutch East Indies to search for *Pithecanthropus alalus*. He gave up his university job, uprooted his

Java Man – Java, in the Indonesian volcanic island arc, was a very important site for early human relatives. It was the furthermost point in Southeast Asia reached by *Homo erectus* around 1.5 million years ago. The fossil evidence suggests that they survived here until at least 100,000 years ago. The dwarfed human related species *Homo floresiensis* of the Indonesian island of Flores may be a descendent of species of *Homo erectus*.

Haeckel's analysis of ethnic differences between modern humans was based on a crude comparison of skull shape and skin colour with an underlying racialist agenda in which white Caucasians and Indo-Aryans were seen as the most advanced and all other populations in a descending scale below these groups. From mapping the distribution of these ethnic groups he concluded that the origin of humanity must have been somewhere between Africa and India perhaps on a now sunken landmass he called 'Lemuria' or 'Paradise'.

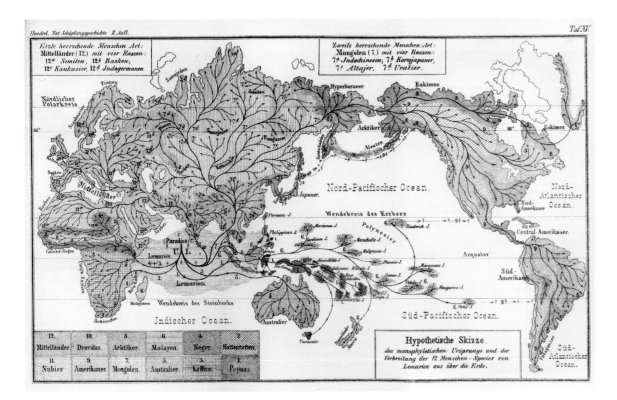

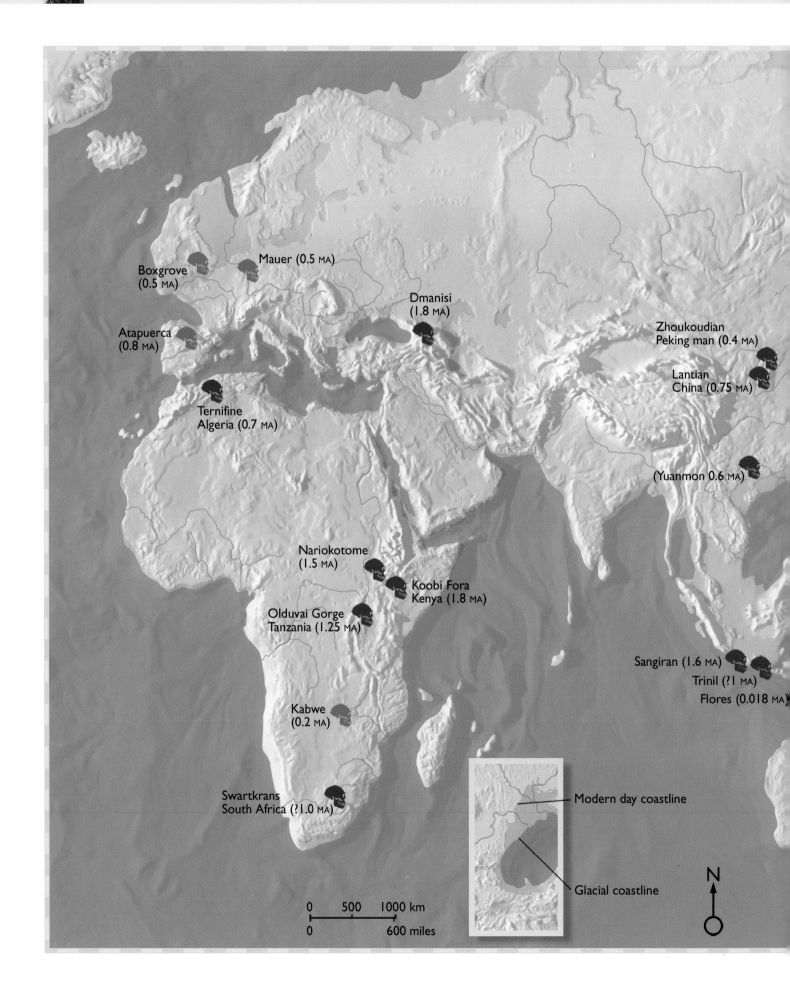

Boxgrove
(0.5 MA)

Mauer (0.5 MA)

Dmanisi
(1.8 MA)

Zhoukoudian
Peking man (0.4 MA)

Atapuerca
(0.8 MA)

Lantian
China (0.75 MA)

Ternifine
Algeria (0.7 MA)

(Yuanmon 0.6 MA)

Nariokotome
(1.5 MA)

Koobi Fora
Kenya (1.8 MA)

Olduvai Gorge
Tanzania (1.25 MA)

Sangiran (1.6 MA)

Trinil (?1 MA)

Flores (0.018 MA)

Kabwe
(0.2 MA)

Swartkrans
South Africa (?1.0 MA)

Modern day coastline

Glacial coastline

0   500   1000 km

0   600 miles

N

H. erectus/ergaster

H. heidelbergensis

H. floresiensis

First out of Africa

The First Out of Africa theory claims that the first human relatives to achieve a widespread distribution were the *Homo ergaster/erectus* people. *Homo erectus* was first recognised as a separate fossil species in Java at the end of the 19th century and subsequently found in China. At the time, it seemed possible that the species might have originated in Asia but there antecedents were not known nor how old they were in chronological terms.

With the discovery of more primitive australopithecine human relatives in Africa, the focus of human origins shifted to Africa. When some *Homo erectus*-like fossils were found in Africa, there was a reluctance to place them in an Asian species and a separate African species, *Homo ergaster*, was created. Now *Homo ergaster* is seen as an African *Homo erectus*, and they are often referred to as *Homo ergaster/erectus*.

One interpretation of the fossil evidence suggests that around two million years ago, at the beginning of the Ice Age, these people, who were only equipped with relatively simple stone tools, found their way out of Africa. There are *Homo erectus*-like fossils at Dmanisi in Georgia, dated at around 1.8 million years old and, in the Indonesian island of Java, *Homo erectus* fossils that are some 1.6 million years old along with further, somewhat younger, finds in China.

However, questions are being asked about the identity of the Dmanisi finds which look more primitive and smaller brained than typical *Homo erectus*. It is suggested that they are more like another African species *Homo habilis*. And, there are new finds emerging from China that may be as old as the oldest African *Homo ergaster* fossils.

Recently, an astonishingly small brained and dwarf human species, *Homo floresiensis*, has been found on the Indonesian island of Flores which may have descended from *Homo erectus* and survived until as recently as 18,000 years ago.

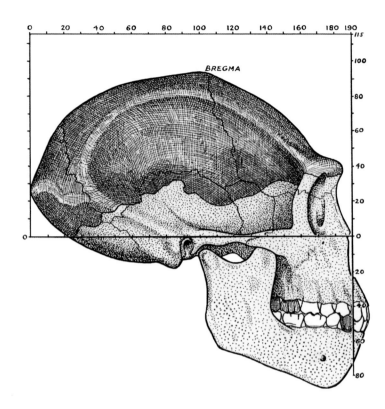

A reconstruction of 'Java Man's' skull from the 1930s in which the preserved fossil remains (the skull cap and a few teeth) have a dark shading. The Neanderthal-like prominent browridge is evident but the facial structure was conjectural. However, the reconstruction is now known to be remarkably prescient.

wife and child for a new job in the Dutch colonial army medical corps in Sumatra but lost no time in persuading the colonial authorities to allow him the time and resources, in the form of two young military engineer assistants and some 50 enforced local labourers, to start investigating promising cave deposits. The results were initially disappointing, however, with only the teeth of relatively recent animals being uncovered. By 1889 he moved on to Java where the recent discovery of a fossil human skull suggested a greater potential. The skull had been sent to Dubois and he knew that it was much more recent than the kind of thing he was seeking, but at least is showed that human-related bones could be found.

By the time Dubois began his search in Java he realized that he needed to look beyond the confines of caves. Abundant fossils of Tertiary animals such as extinct relatives of hippos, rhinos, elephants, cattle, deer and big cats were being found within ancient river gravels, especially in the banks of the modern Solo River. Processing the sediments required an even bigger labour force, which the colonial authorities duly provided, and in August 1891, one of Dubois' engineers found a fossil cheek tooth (molar) and in October a skull cap among various animal bones at Trinil.

On first examination Dubois thought that they were the fossil remains of some kind of chimpanzee but he had no chimpanzee skeletal material for comparison. So, although he realized that he had a potentially important find, at first he adopted the genus name *Anthropopithecus*, which had already been used to describe another fossil primate from the Siwalik Hills in India.

While Dubois was awaiting the arrival of a chimpanzee skull another bone turned up at Trinil in May 1892, and this time it was an almost complete left thighbone. From the reasonably close proximity of the finds, Dubois thought that they had all belonged to the same individual, an assumption that would not be made today. The form of the long, straight robust thighbone convinced him that, as he wrote in 1892, this '...being was... in no way equipped to climb trees in the manner of the chimpanzee, the gorilla, and the orang-utan' on the contrary, it is obvious 'that this bone fulfilled the same mechanical role as in the human body so 'one can say with absolute certainty the *Anthropopithecus* of Java stood upright and moved like a human.' Furthermore, this was proof to Dubois that '...the East Indies was the cradle of humankind'. Apparently Haeckel had been right.

Dubois began to try and estimate the original brain size which was a very difficult task considering that he only had the skullcap. Nevertheless he thought it some 2.4 times bigger than a chimpanzee's and taking a chimpanzee skull volume of 410cc, he calculated that the new skull had a volume of 984cc while commenting that it might well have been 1,000cc or more. With such a large brain, the name *Anthropopithecus* did not seem so appropriate and instead he substituted Haeckel's speculative genus *Pithecanthropus* so the man-ape became the ape-man and he added the species name *erectus*, referring to his claim that it could walk upright. Excavations at the site were renewed with intensified vigour but only one human-related and worn molar was recognized among more animal bones after another year's work.

By January 1894 Dubois had prepared his scientific description of his Java ape-man, which he felt sure was going to stake his claim to scientific fame. He incorporated photographs of the bones and drawings, including comparisons with chimpanzee and gibbon skulls, and concluded that *Pithecanthropus erectus* was the transitional form linking the man-apes with man. But he failed to make a detailed comparison of the find with what was known about the Neanderthals or with the early human fossils known by that time. The find was relatively dated as late Pliocene or early Pleistocene on the identification and correlation of the accompanying animal fossils and the apparent absence of older Miocene ones.

Dubois dealt with the geological context of the find only briefly, as he was by training an anatomist and not a palaeontologist or geologist. In addition, his records of the exact location of the finds were very superficial and unsatisfactory and in any event, he had not been present when the finds were made. Moreover, seemingly he had not made much of an effort to subsequently redress the problem and this was exacerbated by the continuing search, which removed all possible evidence of the context in which the discovery occurred.

Considering all of these factors, he was extremely unwise to claim that the skeletal elements all belonged to the same individual.

Dubois' main interest beyond the basic description was to place his find within an evolutionary context and especially its role in human origins. He knew that true humans were already present in the last interglacial period of the Pleistocene and the implication was that if his Java Man was ancestral to modern humans then development of modern humans must have been very rapid. He got around this problem by describing 'this development of Man *per saltum*' (by jumps), a mode of evolution later promoted by Stephen Jay Gould (1941-2002), the well-known American palaeontologist.

Dubois' results were published in a Dutch academic journal and attracted a fair bit of attention; by the end of the century it was mentioned in some 80 scientific publications. But there was a problem – most of the comments were highly critical. Dubois returned to Holland with his fossils in 1894 hoping to promote his find among the anthropological elite of the day. He tried his best by showing the fossils to his critics and answering their criticisms point by point but faced with almost unrelenting hostility he finally shut the

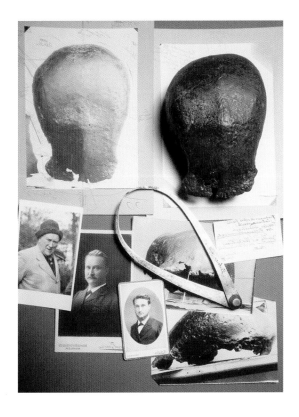

A photomontage shows Eugene Dubois at various stages of his career and part of his most famous fossil find – the skull cap of 'Java Man'. Dubois named it as *Pithecanthropus erectus* and regarded it as the missing link between the apes and humans. But it is now recognised as a member of our genus as *Homo erectus*.

fossils away in 1900. As a last effort, he made a full-size reconstruction for the World Exhibition in Paris in 1900.

His critics picked on many different aspects of Dubois' account, particularly the lack of detailed context and the claim that the remains belonged to the same individual. Many experts thought that the femur was more likely to be human and not related to the skullcap at all, thus undermining any suggestion that the find could represent some transitional form. Some thought that the skullcap came from a large gibbon, others picked on Dubois' lack of comparison with what were then regarded as primitive modern humans or with the Neanderthals which, having a larger brain size, were perhaps more advanced.

At best, a number of experts did admit that it was at least an important find. One major figure who did support Dubois was Ernst Haeckel, who in an 1895 discussion of primate fossils, acknowledged that 'some of these are certainly of great importance, especially the skullcap of the Pliocene *Pithecanthropus erectus* of Java (1894), which really seems to represent the 'missing link' so eagerly sought for, in the chain of transitional forms'. One positive result of all the discussion and argument was that the find became well known and for the first time forced an open

Dubois was unwise to make some of the claims that he did make for his Java ape-man.

Dubois reconstructed his Java Man for the 1900 World Exhibition in Paris as an upright walking being capable of making tools but with ape-like feet and a relatively small brain. Dubois only had a leg bone to go on but was correct in his interpretation of the bipedal stance. He was also a pioneer in the estimation of brainsize from fragmentary skull remains.

debate about human evolution, about what could and could not be considered as human-related and what was ape-related.

The real problem with the Java find was the apparent discrepancy between the humanness of the limb bone and the much more ape-like condition of the skull cap – how could they possibly have belonged to the same individual?

The other benefit flowing from the open debate was the reassessment of the Neanderthal fossils. There was a realization that they were more advanced than the fossils recovered from Java and were therefore perhaps ancestral to modern humans.

Dubois was forced to rethink some of his ideas but he was very stubborn, took the criticism personally and disengaged himself from the debate for some 20 years. He did not even open the many boxes of animal fossils that he brought back from the site in Java. And he refused to provide even casts of the Java find for foreign experts to study.

Eventually, however, in 1922, a complaint from the internationally renowned American palaeontologist Henry Fairfield Osborn (1857-1935) to the Royal Dutch Academy of Science prompted them to intervene. This academic authority at last managed to force Dubois to provide casts and do something with this important collection, but he still prevaricated over producing the results. Eventually, when the boxes were finally opened and catalogued in the 1930s, four more human-related bones were found among the 11,284 animal bones. If Dubois had only bothered to look himself he would have had a much stronger case, although he would have had to admit that his original finds did not necessarily belong to just one individual. By this time, however, other events had overtaken the importance of Java Man.

Curiously, the discovery of such human-related fossils in Java did not lead to much further search in the region, although in 1907-8 another Dutch team did go to Trinil and found many more animal fossils, but no new human-related ones. The discovery of the Heidelberg jaw in Germany in 1907 (see below) and well-preserved Neanderthal skeletons at La Chapelle-aux-Saints (see page 94) brought the focus back to Europe. However, the idea that Asia was the cradle of humanity was still sufficiently strong to spur Roy Chapman Andrews (1884-1960) of the American Museum of Natural History to try searching in Central Asia. He did not find any human fossils but instead some wonderful and amazing dinosaur fossils including nests with eggs and young, but that is another story.

Then in 1908 the 'discovery' of the now notorious Piltdown Man forgery in the south of England meant that most European experts lost any interest in looking elsewhere for human ancestry. However, a new chapter in the story of *Pithecanthropus erectus* was opened up by excavations in the late 1920s carried out at Zhoukoudian (Chou-k'ou-tien) near Peking (today's Beijing) in China.

Peking Man – *Sinanthropus pekinensis*

Chinese traditional medicine uses many strange ingredients, including so-called 'dragon bones and teeth' which are ground into a powder as a cure for certain ailments. Over millennia fossil bones and teeth from sources as diverse as dinosaurs and primates have been excavated purely for this use and sold in rural markets all over this vast country, and some localities have become well known as good sources. Dragon Bone Hill at Zhoukoudian, with its limestone caves, was just one of these sources that attracted the attentions of an international team led by Canadian anthropologist Davidson Black (1884-1934). After studying with Sir Grafton Elliot Smith (1871-1937) in London, Black had taken up a job in Peking Union Medical College in the hope of finding ancient human-related fossils in China. He eventually found some promising looking teeth, which helped persuade the Rockefeller Foundation to finance more systematic excavation in 1927. This yielded a new tooth, which he thought warranted a new genus and species called *Sinanthropus pekinensis*.

Nowadays, a single tooth would not be nearly enough to warrant the naming of a whole species, let alone a new genus, but it guaranteed Black more funding. It took another two years before Black's Chinese colleague W.C. Pei (1904-1982) made a startling discovery. On December 2nd, 1929 he uncovered a well-preserved brain case, but without the facial bones. By the end of the month the news had spread worldwide, the era of the press conference and international telegraph made sure of that. The hubbub also obscured attempts by Raymond Dart (1893-1988) to get some attention paid to his new African find of the Taung 'child', *Australopithecus africanus*. By comparison with the Asian material, the little African skull, however complete and beautiful, seemed to the experts of the day to be far too primitive to justify much attention.

Black concluded that the skull, along with another one found in 1931, belonged to his new Peking Man, *Sinanthropus pekinensis*. With brain sizes of 918cc and 1150cc, they showed features which placed Peking Man at the base of a Y-shaped evolutionary branch in which one spur led to the Neanderthals and the other to modern humans. The German anthropologist Franz Weidenreich (1873-1948) agreed with Black but thought that there were

sufficient similarities between Peking Man and Java Man to justify grouping them together. Some other experts, notably Marcellin Boule (1861-1942) in France, agreed that there should be one genus *Pithecanthropus* because it had priority with two species and that the genus name *Sinanthropus* be dropped.

After Black died in 1934 Weidenreich took over supervision of the excavations and in 1937 they had a bumper 'catch' of five skulls with brain sizes ranging from 850cc to 1200cc, thus bracketing Java Man's 984cc brain and emphasizing the similarity of these Chinese and Javanese human relatives. The great advantage of the Chinese finds was that they included some amazingly complete skulls – some of the best preserved of our ancient extinct relatives ever found. They have the now-familiar prominent bony brow-ridges, backward-sloping 'forehead', forward-projecting 'muzzle' and receding chin, all rather like the Neanderthals. However, the face is relatively smaller and somewhat more modern looking, as are the proportions of the limbs and build of the skeleton.

Meanwhile, W. F. F. Oppenoorth, a mining engineer with the Dutch colonial geological survey, found more human-related fossils in the Solo River valley near Ngandong in Java and by 1933 they had 11 partial skulls (and two fragments of leg bones) among thousands of animal bones. From these relics it was deduced that they represented a distinct species, which was duly dubbed Solo Man. Dubois

Peking Man – Although originally thought to be a separate species, the ancient fossil human relatives of China, that date back to nearly two million years ago, are now known to belong to *Homo erectus*, the same species as 'Java Man'. Some of the best preserved ancient human related remains, many of the finds were lost during the civil unrest of the 1900s.

With the advent of the news conference and telegraph, news of new discoveries and claims now quickly reached a world-wide audience.

The richness and high quality of the Chinese skeletal remains from Zhoukoudian near Beijing surprised the world of anthropology. This photomontage shows Conrad Black who led the excavations and some of the skull material the team recovered. Their excavation required a great deal of time and effort but the fossils were some of the best known. Tragically, most were lost during the chaos that followed the Japanese invasion of China in 1940. The only compensations is that some high quality casts were made and did survive shipment to America.

The 1940 invasion of China by Japanese forces led to the loss of precious fossils.

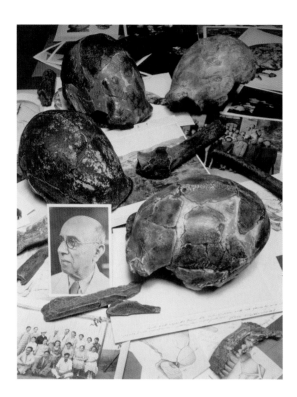

calculated that the brain size was 1,050cc for the females and 1200cc for the males, which he claimed was lower than that of the Neanderthals but higher than that of some modern native Australians. Solo Man was even given a new species name by Oppenoorth and placed in the genus *Homo* as *H. soloensis*. He also suggested that it was possible to recognize two geographically separate lineages, a European one of *heidelbergensis* → *neanderthalensis* → *sapiens* and an Asian one of *Pithecanthropus* → *Sinanthropus* → *Homo soloensis*. This was an idea that was subsequently taken up and transformed into the 'Multi-regional Hypothesis', becoming the subject of an ongoing bitter argument for decades.

Between 1936 and 1940 more skull and jaw fragments were found by Oppenoorth's successor in the Java geological survey, an ambitious young German geologist by the name of G.H.R. (Ralph) von Koenigswald (1902-1982) at Modjokerto and Sangiran in Java. Again, however, there were problems with the exact location of some of the finds, which had been made by local helpers. The Sangiran braincase was found in fragments and Von Koenigswald had to reconstruct the skull before he could make any estimate of the brain size – which turned out to be surprisingly small at 775cc. Nevertheless, it also showed some very human features in the jaw articulation and position of the

ear opening and Von Koenigswald referred to it as *Pithecanthropus*, much to Dubois' annoyance. Early in 1939 Von Koenigswald took his new Java finds to Peking where Franz Weidenreich had laid out the Chinese and Javanese specimens side by side for direct comparison.

The Chinese material was light in colour and much closer to the original bone colour because it was not nearly so heavily mineralized as the dark-coloured Javanese specimens, which had mostly been buried in ash deposits from the numerous highly active volcanoes of the Indonesian archipelago. The Javanese fossils are generally more robust, with thicker skulls but smaller brains than the Chinese ones. But as Von Koenigswald later wrote '...in every respect they showed a considerable degree of correspondence', a conclusion they echoed in a subsequent paper to *Nature* (1939, vol. 144, pp. 494-6). Weidenreich went on to develop the idea that the Java Man subsequently evolved into Australasian modern humans while Peking Man evolved into eastern Asian modern humans and the Neanderthals became modern humans in Eurasia, an idea that became known as the Multi-regional hypothesis of human origins. Few now support the idea, however, which flies in the face of the genetic evidence for the homogeneity and recency of modern humans.

In 1940 the Japanese invaded China and Weidenreich had to flee to America, but such was his haste that there was no time to take the fossils with him. They were left to be crated up and shipped on later and it is one of the great tragedies of palaeoanthropology that only one crate survived to reach America, and it was full of casts. Nobody knows exactly what happened but the generally accepted story is that the crates were sent to Shanghai for shipment, but how many of them were actually put on board is not known; the missing crates may have been lost or stolen *en route*.

Back in Java, Von Koenigswald was interned by the Japanese but later released and got to America while the Javanese fossils were safely hidden away in Java and survived the war. Dubois survived just long enough to see his homeland invaded by Hitler's forces.

In 1942 Ernst Mayr (1904-2005), the American evolutionist, reassessed the whole question of

Pithecanthropus and *Sinanthropus* and their relationship to *Homo*. He concluded that the Asian fossils belonged to a single taxon so closely related to *Homo* that the species should be included within it. Since Dubois' *erectus* had priority, *pekinensis* and *soloensis* were suppressed and they have all become known as *Homo erectus*. Then, when it became possible to radiometrically date the finds there were some surprises, but because Dubois had not involved himself in detailed examination of the animal fossils which had accompanied his original find, he had very little idea about their relative age.

We now know that the oldest Javanese specimens are around 1.5 million years old (early Pleistocene in age) while the youngest may be as little as 100,000 or even 50,000 years old. For many years the latter date seemed incredible because it implied that *Homo erectus* people might still have been living in the Indonesian islands when the first modern humans arrived. But the recent discovery of the fossil remains of the little *Homo floresiensis* people on the Indonesian island of Flores amply supports the idea. By comparison, the Chinese specimens seem to have a smaller age range from around 1 million to 250,000 years. The big question was: where did *Homo erectus* originate?

The origins of *Homo erectus*

The first find beyond the Far East of fossil bones which compared with *Homo erectus* was made by the Leakeys in Olduvai (see Chapter 5). It was a fragmentary skullcap dated to around 1.2 million years but since then older finds have been made in northern Kenya; these include a skull dating to

between 1.7-1.8 million years. The most spectacular find however was the Nariokotome boy – the virtually complete 1.5-million-year-old skeleton of a young boy of between 8 and 11 years old when he died (see p. 131).

The question here is whether or not a single species could range from East Africa to the Far East

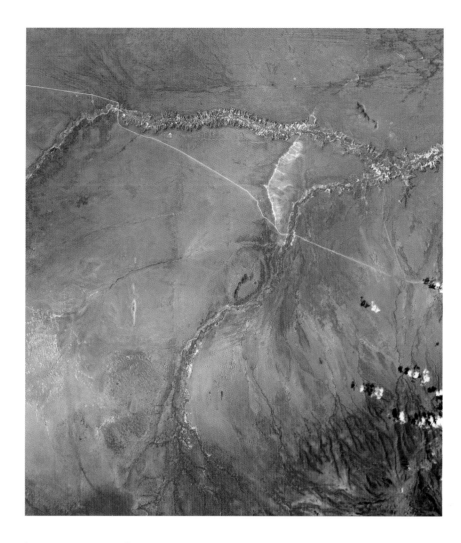

This satellite image of Olduvai Gorge in Tanzania shows how the gorge has been incised into the Serengeti Plain with the Ngorongoro volcanic crater just visible in the bottom right hand corner.

In 2003 fossil footprints, some 350,000 years old, were discovered imprinted in volcanic ash within the Roccamonfina volcanic complex in southern Italy. They are the oldest known tracks made by a member of the genus *Homo* and probably belong to the species *Homo heidelbergensis*.

Some of the oldest finds outside of Africa have been made in Caucasian Georgia.

over a period of nearly two million years and still be the same species, especially given the relatively small size of the overall populations and their likely reproductive separation from one another by geographical and climatic barriers. In other words, should the African specimens be included in *Homo erectus* to emphasize their evolutionary connectedness or, distinguished separately on more realistic biogeographic grounds? Not surprisingly, experts differ on the subject so that the Nariokotome boy is seen as an example of African *Homo erectus* by some and as a member of a separate African species *Homo ergaster* by others. What we have to remember here is that fossil species are essentially the constructs of scientific opinion and different from biological ones since the reproductive test for a species cannot be applied to fossils. The same problem has arisen with some spectacular finds made in recent decades from Georgia in southern Eurasia.

Dmanisi

Several skulls and other skeletal bones have recently been recovered from excavations below the mediaeval village of Dmanisi in Caucasian Georgia – they are some of the oldest human-related remains to be found outside the African continent. Around 1.8 million years ago at the beginning of Pleistocene times the area was located on a peninsula between the Black and Caspian Seas and was part of an

Dmanisi – Skulls, bones and tools, some 1.8 million years old, found at Dmanisi in Caucasian Georgia have caused quite a stir. The skulls are strangely varied and do not fit easily into *Homo erectus*, as expected but may belong to some more primitive species.

important land corridor connecting Africa and the Middle East with Eurasia.

A succession of remarkable finds has been made by Georgian palaeontologist David Lordkipanidze (1963-) and his colleagues since 1991. The first find was that of a large jawbone and it was followed by bones from at least six individuals, including four skulls of a *Homo erectus*-like species. Initially the finds were seen as belonging within the known variation of *Homo erectus* but some of the skulls have a very small brain size, a small nose, narrow brow-ridges and large canine teeth. There is a suggestion that they are strongly dimorphic and closer to the African species *Homo habilis* and would be better placed in yet another species *Homo georgicus*. However, with an age of 1.8 million years they predate the classic Asian *Homo erectus* and are virtually contemporaneous with the African *Homo ergaster/erectus*, so perhaps it is not surprising that they retain some primitive features.

One of the fascinating discoveries is that of a virtually toothless skull and lower jaw with resorption of the bone following the loss of all but one of the teeth. It seems that this particular elderly individual continued to live for some time after the teeth had been lost. Clearly, survival for that individual depended upon a supply of soft food and probably the support of others. Until now the archaeological evidence for such 'support in the community' has been largely restricted to modern humans and perhaps some Neanderthals. Such a caring tendency equates with a conscious ability to imagine or see beyond the self and have a sense of the 'other', a sense that becomes formalized in ritual burial. The assumption has been that prior to the development of such a sense the sick and elderly were normally abandoned to their fate even by their 'nearest and dearest'. Old age began before 40 years and most individuals were dead by the time they were 45 years old, which is of course old by most mammal standards – but not all. Famously, elephants can survive into their 30s and 40s even in the wild.

The human-related fossils at Dmanisi are accompanied by thousands of simple stone tools, especially simple choppers and scrapers, along with hundreds of bones of animals such as giant deer, wolves and sabre-toothed cats. Altogether, they show that this was a favoured site for these early human

Dmanisi

RUSSIA

CAUCASUS MOUNTAINS

Black Sea

GEORGIA   TBILISI

Dmanisi
Homo erectus
1.8 ma
stone tools

Caspian Sea

N

TURKEY

0   100 km

relatives. But their habitual presence at the site also drew the attention of predators such as the sabre-tooth cats – paired holes on one of the human related skulls have the same spacing as the fangs of a sabre-tooth. Like many big cats they dragged their prey by the head or neck to where they could not be bothered by scavengers such as wolves.

There is still much information to be recovered from Dmanisi but overall it supports the Out of Africa theory for the dispersion of an African *ergaster*-like population that dispersed and evolved into Asian *Homo erectus*. Meanwhile, back in Africa and Eurasia, human evolution continued and produced yet another species – *Homo heidelbergensis*.

Heidelberg Man

The origin of the name Heidelberg Man can be traced back to 1907 and the discovery of a well preserved and massive lower jaw in a sandpit at Mauer, near Heidelberg in Germany. The teeth are distinctly human in appearance but the jawbone is very thick, well developed and lacks any external bony chin process. So it has some similarities to a Neanderthal jaw but is much more robust. As was the tendency at the time, it was soon given a separate species name (*Homo heidelbergensis*), although no other remains of the skeleton were found with it. Because of this lack of supporting evidence, for a long time it was just a single curiosity and many experts regarded it as a European representative of the much better known Asian species *Homo erectus*.

We now know that the Heidelberg jaw dates from around 500,000 years ago and is thus significantly older than the oldest known Neanderthal remains. However, the species was subsequently resurrected to incorporate a number of other finds, one of the most important of which came from Africa and was at first known as Rhodesia Man.

The Kabwe skull – Rhodesia Man

Commercial copper mining at Broken Hill (now Kabwe) in what was Northern Rhodesia (present-day Zambia) has been conducted on a massive scale for many decades. In the 1920s miners occasionally found bones covered in mineral growths as they quarried through part of a cave system, and they either discarded the relics or threw them into the furnaces to be smelted with the other ore.

Fortunately, however, the discovery of a fine skull in 1921 by one of the miners, Tom Zwigelaar, was so impressive that it was not discarded. Again it was a new specimen – and as mentioned earlier, scientists decided it was a new species and warranted a new name and so it became Rhodesia Man (*Homo rhodesiensis*) for a while.

The skull is incredibly well preserved and again shows the primitive feature of a very thick brow-ridge that arches over each eye socket (orbit). Although there is virtually no forehead, the face has some advanced modern features and the brain is relatively large and approaches the size of modern humans. In addition an associated limb bone suggests that it was an individual who was tall and well built, and part of the pelvis found with the skull has a distinctive socket that resembles a feature seen in the more primitive *Homo erectus* and *Homo heidelbergensis*, but not in the Neanderthals or us.

As no more specimens were found at the site, the fashion for proliferating new names based on single specimens began to wane and experts preferred to shoehorn Rhodesia Man into the nearest similar and well established taxon – *Homo erectus* since Heidelberg Man was still an equally problematic species. However, over the years the appellation has been attached to a number of other finds in Europe such as a skull from Petralona in Greece, Arago in France, Bilzingsleben in Germany and a thighbone from Boxgrove in southern England.

Rhodesia Man's skull was incredibly well preserved, and has some modern features.

In the 1920s, miners at Broken Hill, in what used to be Northern Rhodesia (now Zambia) occasionally found human-like bones in the limestone caves they were tunnelling through. Most were discarded but a very well preserved skull was saved and recognised as a new species called *Homo rhodesiensis*. Now it is recognised as a member of our directly ancestral species *Homo heidelbergensis*.

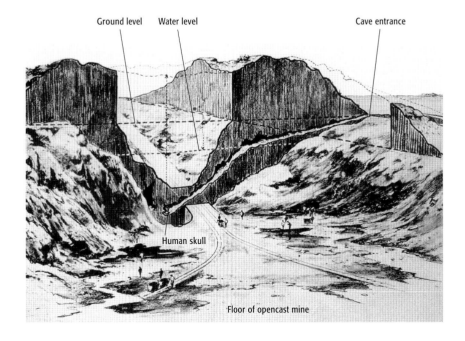

Ground level   Water level   Cave entrance

Human skull

Floor of opencast mine

The Boxgrove Butchery

Along the south coast of England, Cretaceous age chalk forms a famous line of white cliffs as erosion by the sea cuts into the soft limestone. However, during the Pleistocene ice ages sea-levels were at times significantly higher than present and the sea reached far inland, cutting into the chalk and creating cliffs that are still preserved but are now some kilometres from the present coast. In places younger sands, silts and gravels were deposited on top of the beaches cut in the chalk and are now exploited commercially.

Pakefield – The recent discovery of 700,000 year old stone tools at Pakefield in East Anglia has pushed back the record of human occupation of the British Isles by about 200,000 years.

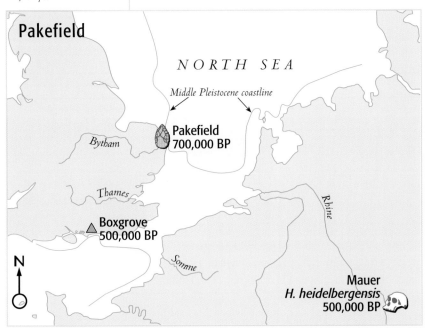

At Boxgrove, near Chichester in West Sussex, such activity in the early 1980s uncovered a remarkable prehistoric surface, which had been used as a prehistoric butchery site at the base of a huge chalk cliff around 500,000 years ago – long before the Neanderthals occupied western Eurasia. Two phases of excavation led by Mark Roberts between 1982-6 and 1993-5 uncovered thousands of stone tools along with hundreds of animal bones and some human-related remains.

Originally, in the early Pleistocene times there was a series of lagoons and beaches below the chalk cliffs and these attracted a host of diverse animals and their predators – both animal and human-related. The sequence of land surfaces at Boxgrove spans some tens of thousands of years and preserves a range of ancient environments, from coastal marine to marshland and open tundra to grassland, so forming a record of rapid climate change. Herds of game grazed the coastal plain with horses, bison, red deer, elephants and rhinoceros followed keenly and closely by their predators – the big cats, hyenas, wolves and some of our human relatives.

The site would have provided an ideal combination of game and high-quality raw material for tool making – flint – in close proximity to each other. The flint occurs naturally in the chalk limestone and is the result of a strange mineral phenomenon (see box on page 118). Some 300 almond and pear-shaped hand axes have been found at Boxgrove, along with the debris of fragments left over from the constant manufacture and renewal of stone tools as animal carcasses were butchered.

By mapping the distribution of the chips and bones it is possible to reconstruct the original scene and activities. For instance, analysis of the remains of one horse shows that it was butchered in seven or more stages with the flint tools being renewed at each stage. Following the initial kill, the animal was gutted and the main marrowbones exposed. The latter were smashed open and the marrow and other soft tissue such as the liver eaten raw on the spot. Only the big muscle blocks and skin were removed and carried away.

Eight rings and piles of flint chippings show that at least eight hunters knelt around the carcass to butcher it and renew their stone tools as they did so with the chips falling between their knees.

Occupying England

Evidence for the earliest occupation of the British Isles has been pushed back to 700,000 years ago with discoveries at Boxgrove in the south of England and more recently at Pakefield in East Anglia. The Boxgrove site has been dated to around 500,000 years ago, similar to the Mauer site in Germany, discovered in 1907, and Pakefield dated to 700,000 years ago. At Pakefield, an Ice Age deposit called the Cromer Forest Bed outcrops in a seacliff from which bones of the extinct animals of the period are washed onto the beach. Some keen amateur collectors also excavate the deposit in the cliff and it was here that over 30 primitive stone tools have recently been found along with bones of small mammals such as hamsters, squirrels and bats that lived in the original interglacial forest. The fossil bones also allow the relative dating of the deposit.

There has been a long-running argument over whether these extinct human relatives were opportunist scavengers who just took advantage of kills by other predators or, whether they actively hunted and killed their own prey. Well, Boxgrove is one of the sites that demonstrates that they were indeed active hunters. Four rhinos were skillfully butchered at this site and these are animals that, once they are mature, have no known regular animal predators. Detailed examination of the butchery marks on the animal bones shows that the animals were initially intact and any gnaw marks from animal scavengers post-date cut marks made by the human-related butchers.

Furthermore, one horse shoulder blade has neat circular perforations that are almost certainly spear wounds and thus strongly indicative of active hunting. A single mature rhino would provide some 700kg (1,540 lb) of food and a horse some 400kg (880 lb). The efficiency of the butchery shows considerable economy, suggesting that either larger groups were being fed and maintained, or that the meat was being treated and stored for later use. Interestingly, there is no sign of fire being made at Boxgrove.

The very careful excavation of the Boxgrove site in the South of England revealed a wealth of detail about life at this location around 500,000 years ago. Animal bones with cut marks show that four rhinos were butchered here with stone tools. A few human remains suggest that it was *Homo heidelbergensis* who occupied the site.

The human-related remains at Boxgrove are very sparse but they're significant. They include a gnawed shin bone and two large and well-worn teeth and represent the oldest human-related skeletal material yet found in the British Isles. The shin bone is that of a mature and well-built individual who probably stood around 1.82m (6 ft) tall and weighed some 82kg (180 lb). The heavy wear on the teeth shows they had been frequently used as grippers for some material such as animal skin and surface scratches caused by accidental contact with stone tools indicate that the individual was right-handed. Pieces of skin were probably held in the teeth as tissue and fat were cut and scraped away, or pieces of flesh were held in the teeth and left hand and then cut

Evidence of crossbreeding

The discovery in the late 1990s of the buried remains of a four-year-old child at the Abrigo do Lagar Velho, Portugal has generated considerable controversy. Dated at some 24,500 years old, the burial with a pierced shell and red ochre has the hallmarks of the Upper Palaeolithic and early modern humans in Europe. The skeleton also shows many early modern features, especially in what remains of the skull and teeth. However, the proportions and robustness of the limbs are more characteristic of the Neanderthals.

The Portugese scientists and their collaborators, who have described the find, argue that this mosaic of features is the result of crossbreeding and hybridisation between a late surviving Neanderthal population in southern Iberia and incoming modern humans. Unfortunately as no DNA has been recovered it is not possible to test the claim for supporting genetic evidence.

A reconstruction of Boxgrove shows *Homo heidelbergensis* butchering a rhino some 500,000 years ago.

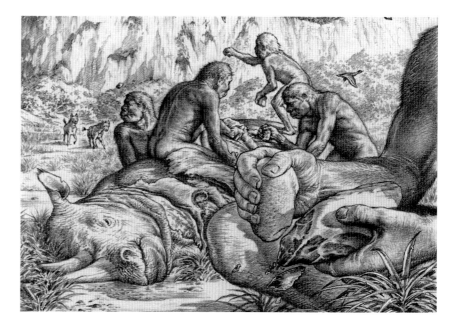

Flint – the earliest cutting edge

Flint (also known as chert) is one of the best, and best known, materials for making stone tools. Technically, it is a cryptocrystalline mineral, meaning it's very finely grained without well-defined crystals (cryptocrystalline rocks are composed of crystals that can be seen only under a polarizing microscope). As a result it behaves rather like glass, fracturing along concave curved surfaces into flakes that can have very sharp edges.

With practice it can be easily fashioned (or knapped), with simple stone and bone tools, into a variety of shapes from large axe heads to tiny razor-sharp shards. Being so brittle, however, the edges are soon blunted with use, but new flakes can be removed a number of times to produce new edges on blades, points and choppers.

Geologically, flint is found in sedimentary strata, especially limestone rich in calcium carbonate - although flint itself is silicon dioxide (SiO_2). The silica was derived mostly from the shells and skeletons of sea-dwelling organisms, especially silica sponges and radiolarian protists (protozoa). When these organisms died their silica-rich skeletons became a part of the surrounding carbonate sediment. This meant that the skeletons were of a very different chemical composition to the material in which they were embedded. So instead of being preserved, they disintegrated.

As they did so, silica-rich fluids were produced within the sediment and as the concentration grew, so these eventually re-precipitated the silica in the form of nodules that grew within the sediment, in an infinite range of shapes, sizes and abundance depending upon the overall amount of silica present. Often the nodules grew around the remains of fossils such as sponges as the soft sediment turned into hard sedimentary rock. The process still continues today, but obviously the flint we're talking about was formed many millions of years ago.

In the northwest European context flints are very abundant in soft limestone (chalk) of the Cretaceous period. The chalk outcrops from Denmark in the north and then to the south through the low-countries, England and France. And, because flint is relatively hard compared with the chalk, the nodules remain when the chalk is eroded and weathered away. Subsequently, the residual flints have formed extensive gravel deposits on beaches and in riverbeds. Often, however, these flints are highly weathered themselves and not as tough as fresh ones – those that have been protected by the chalk covering them.

When fresh, flints are dark grey to black and slightly translucent. They often have to be quarried from the chalk but as the latter is comparatively soft, that can be done with relative ease. Ancient human-related occupants of Europe from *Homo heidelbergensis* onwards, who manufactured stone tools, were well aware of the properties of flint and where to find it. They were prepared to travel considerable distances to recover good-quality stone and transport it in a semi-finished state to wherever it was needed.

---

into smaller, more manageable pieces by a stone tool held in the right hand.

The indications from the skeletal remains suggest that the human-related species here at Boxgrove was *Homo heidelbergensis* but they are not the oldest human-related archaeological remains to be found in the British Isles as some even older stone tools have been found in East Anglia.

The oldest known 'Brits'

The River Thames which flows through London today is a mere shadow of its former self… At times during the Pleistocene ice ages the 'proto-Thames' and its tributaries were mighty rivers carrying huge volumes of glacial meltwater into the North Sea. In doing so they also formed vast sandbanks and terraces of stony shingle on either side of their numerous channels and beyond these were floodplain deposits and marshes which were home to a great diversity of plants and game. These attracted some of the earliest human-related inhabitants of Northern Europe.

Today, it is mostly the sands and gravels of the old river terraces that remain, but in places these contain animal bones and recently stone tools have been found in some of the oldest terrace deposits in East Anglia. These tools have been dated to Cromerian times, around 700,000 years ago, which makes them significantly older than Boxgrove. Unfortunately, however, despite intensive searching no human-related bones have yet been found. Interestingly, these stone tools are also more primitive than those found at Boxgrove in that they only consist of choppers and scrapers, there are no sophisticated hand axes.

So were these tools made by a different human relation other than *Homo heidelbergensis* and if so, do they represent a separate wave of early settlers in Europe? The new English discovery builds on another early find near Burgos in northern Spain at an extraordinary site called Gran Dolina in the Sierra de Atapuerca.

Spanish 'cousins'?

The late Cretaceous age limestone hills of the Sierra de Atapuerca are riddled with caves, tunnels and other structures formed over the millennia by the corrosive action of slightly acidic rainwater. Naturally, many of these caves and tunnels would have remained undisturbed, but between 1896 and 1910 the construction of a 65km (40 mile) railway line involved extensive cuttings and tunnels through the Sierra. In the process a number of subterranean caves and passages were exposed in the walls of the

Excavation within the limestone caves and passages of the Sierra de Atapuerca in northern Spain has uncovered a wealth of fossil remains that are helping to fill a few of the many gaps in the early occupation of Europe by our extinct human relatives. The oldest remains, some 780,000 years old, come from the site known as Gran Dolina. Here, a number of disarticulated skeletons, mostly of children, are covered in cut marks evidently made apparently with intent by their fellow beings with stone tools but for some unknown purpose.

cuttings and modern excavation of one of these ancient chambers called Gran Dolina has revealed a wonderful assemblage of human related bones buried within the cave fill sediments.

The total assemblage includes the disarticulated remains of several individuals, mostly children — and notably the bones are covered in cut marks. The relics included backbones, ribs, limb and foot bones along with part of a skullcap and facial bones with some teeth still in place. Microscopic analysis shows that stone tools rather than the teeth of scavengers made the cuts. The stone tools found with the bones are fairly primitive blades and choppers of a pre-Acheulian type. This 'butchery' could be the result of burial practices, which included dismemberment and defleshing or possibly cannibalism.

Dating of the finds initially relied on the discovery of many small mammal bones, which included species that died out over 500,000 years ago. But then palaeomagnetic measurements (see box on palaeomagnetism) of the cave fill sediments revealed the presence of a switch in the Earth's magnetic field, which is known to have occurred around 780,000 years ago at the beginning of the present Brunhes magnetochron.

The skull shows a mixture of primitive features similar to those seen in some African fossils over 1.5 million years old. However, the forehead with its well-developed and arched brow-ridge suggests that the child had a rather larger brain than typically found in African *Homo erectus/ergaster*. The Spanish

anthropologists who excavated the find believe that the relic shows enough new features to warrant a new species, which they have called *Homo antecessor*, meaning 'pioneer man'. They claim that it originated in Africa and spread to Europe where it subsequently evolved into the Neanderthals. If this were true then the split between the Neanderthals and modern humans would be very ancient, and this is something that is not supported by the recovery of Neanderthal DNA. In addition, they claim that its African population also gave rise to modern humans — our very own *Homo sapiens*.

As we can seen, over recent decades evidence has grown of a far from simple picture of human

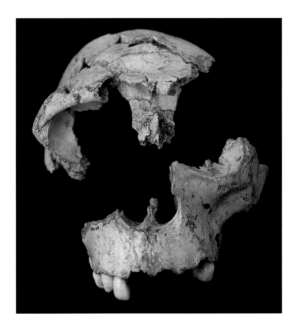

A fragmentary skull from Gran Dolina suggests that this population was close to *Homo erectus* but the Spanish experts have named them as belonging to a new species *Homo antecessor*.

The evidence that has been gathered now points to a far from simple picture.

Palaeomagnetism – using magnetism to plot the continents' drift and date fossils

Some iron-rich mineral grains act as natural compass needles and become oriented parallel to the Earth's prevailing magnetic field at the time of their deposition or formation. Thus the orientation of that field is 'locked' into the rock record in perpetuity provided it is not reset by reheating. The parent sediments and rocks can be buried, folded and even moved around the Earth's surface by plate tectonic movements and yet the rocks will retain that original magnetism and its geometric co-ordinates relative to the original field. This property of certain rocks to retain a record of their original magnetic orientation has proved invaluable to geologists and archaeologists for a number of reasons but especially the reconstruction of past movements of the Earth's crustal

plates and the continents carried by them.

The Earth's magnetic field approximates to that of a bar magnet (known as a simple dipole field) with two poles, today's North and South. They lie on an axis that is roughly similar to that of the Earth's axis of rotation (geographic north) and oscillate within a few degrees of it in an irregular fashion. Even less predictable is the tendency for the Earth's magnetic field to 'flip' (reverse) its polarity over a relatively short period of time (a few thousand years) so that north becomes south. A geological record of these magnetic reversals is contained in the stratigraphic succession of ancient sediments and interbedded igneous rocks that have accumulated to form the Earth's crust over the last three billion and more years.

Recently geologists have been able to recover the chronology of past reversals of the magnetic field from the stratigraphic rock record and construct what is known as a magnetostratigraphic correlation and relate it to the established stratigraphic record based upon sedimentary rocks and their contained fossils. Since the development of radiometric dating it has been possible to put dates to certain points in this record. So we now have a chronology that is calibrated in years before the present and one that includes the history of magnetic reversals.

For instance the current normal field, called the Brunhes magnetochron (after the French geologist Bernard Brunhes who first discovered the reversals in the magnetic field) has held over the last 0.78 million years. Before that there was the

shorter, 0.2 million-year-long Matuyama reversal from 0.99 million years ago and so on.

Identification of reversals in sedimentary sequences can help pin down time intervals and put constraints on palaeoanthropological materials. For instance, the Olduvai Gorge succession in Tanzania of sediments and volcanic materials includes a record of a normal phase of polarity known as the Olduvai Event, which we now know from radiometric dating lasted 0.17 million years from 1.95 million years ago. Identification of this short phase of normal magnetism within an otherwise much longer phase of reversed polarity is of considerable value for the dating of sediments and fossils in East Africa and beyond because it coincides with the base of the Pleistocene epoch of geological time.

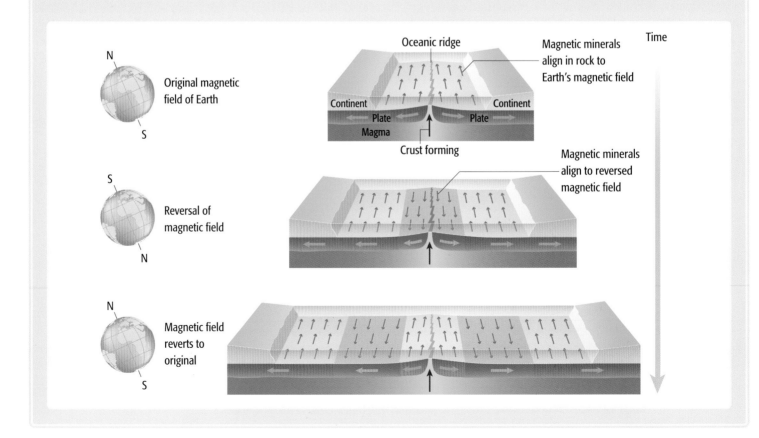

migration and evolution 'Out of Africa'. The scattered European finds, which span some 800,000 years, complicate a simple story of just two moves out of Africa. There is no doubt that there was an early 'diaspora' nearly 2 million years ago by *Homo erectus/ergaster* and that they got as far as Southeast Asia where their dwarfed *Homo floresiensis* descendants may have survived until as recently as 19,000 years ago. By then they would probably have encountered modern humans as they moved through the Indonesian archipelago and onwards to Australia.

However, the emerging European story suggests that there was another later radiation just under a million years ago by *Homo erectus/ergaster* or some close relative such as *Homo antecessor* out of Africa and into Europe where they perhaps evolved into the Neanderthals. But where does that leave *Homo heidelbergensis*, which has also been seen as a possible ancestor to both the Neanderthals and, back in Africa, to our species? Clearly one hypothesis is wrong but it is not entirely clear which one, although at the moment the *heidelbergensis* theory probably has more support in the anthropological community as a whole.

Support for an Asian homeland for native Americans comes from archaeological, anthropological and genetic studies. Genetic analysis of living native American populations verifies their link to modern Siberians, although in detail the story may well be complex. There is emerging evidence of more than one phase of migration into the Americas and replacement of earlier populations by more recent ones – which is why finds such as that of Kennewick Man are potentially so important (see box).

There is archaeological evidence from Siberia showing that modern humans arrived in the region as long as 28,000 years ago. Their stone tool technology included the manufacture of tiny blades, called microliths, which greatly improve the penetrative power of wooden spears when entering the flanks of fur-clad animals. It is thought that these early Siberians provided the population from which successive small groups of mobile hunters first migrated into the Americas as they followed the game on which they depended for food and many other necessities of life.

Kennewick Man

The well-preserved skeleton of a middle-aged man was found on July 31st, 1996 on the bank of the Columbia River at Kennewick in Washington State, USA. It had been washed out of the riverbank sediments and this apparently innocuous discovery has since proved to be enormously controversial. When examined in detail the skull shows certain non-Paleoindian features, which some people claim are Caucasoid. Then, when it was dated to around 9300 years, the find became the focus of a great deal of attention with all sorts of claims being made about its possible origin. These could have been resolved by further tests and analysis especially the retrieval of DNA from the bones. However, the whole issue became complicated when the remains were claimed by four Native American tribes as those of an ancestor who ought to be reburied rather than be studied by archaeologists.

As a result, the US Army Corps of Engineers handed the remains over to the tribes on whose land they had been found. The scientists, who were trying to discover the origin of this important find, sued for the right to study the remains. They won their case but not before it went to the San Francisco-based Ninth Circuit Court of Appeals who also rejected an attempt by the tribes to have the decision reviewed by the US Supreme Court. The ruling said that it was impossible to establish a relationship between the Indian tribes and the remains as was claimed.

However, the scientists are now fighting for the right to obtain DNA from the remains as this would be the only way to learn more about Kennewick Man's true relationships. Meanwhile, the skeletal remains consisting of 380 bones and bone fragments are now stored in the Burke Museum in Seattle.

We already know that Kennewick Man led an eventful life. When just a teenager he suffered a near fatal stab in the hip – being struck with such force that the stone tip broke off and was left embedded in his pelvis bone. Remarkably, he survived. The bone grew over the stone point and equally remarkably, he did not suffer from arthritis and died aged between 45 and 55 years old. This was normal life expectancy for the time and way of life. When he died, his bones must have been covered up almost immediately – before any scavenging animal could gnaw at the cadaver or carry off any part. His body may have been naturally covered by flood-borne river sediment or some other natural event, but it is also possible that his fellows, who lived on the rich hunting and fishing land around the confluence of the Columbia and Snake Rivers, buried him.

Despite certain Caucasoid features, Kennewick Man's ancestors were almost certainly Asian and part of the initial movement of people from northeastern Asia who gradually crossed the Bering Land Bridge or paddled along its shoreline when the land bridge was exposed, thousands of years before their descendants lived along the Columbia River. Other relatives of these same distant ancestors of Kennewick Man moved south into what is now Japan, coastal China, and onto the islands of the Pacific.

CHAPTER FIVE

Searching for the origins of humanness (defining the 'floor' to humanness)

Some 20 of so different human-related species have been described so far — and there are more hidden away and awaiting description and categorization in the palaeoanthropological laboratories of the world. Of this total, six are generally accepted as members of our genus *Homo*, 12 belong to australopithecine and older species and at least two (*habilis* and *rudolfensis*) lie in a disputed 'no-mans-land' between the two major groupings. So why should it be so difficult to define the basic attributes of our genus?

This chimpanzee is using a stone to crack open walnuts. The use of tools is extremely rare in animals, and is seen as a sign of intelligence. It has been shown that chimpanzees' learning abilities are similar to those of young children.

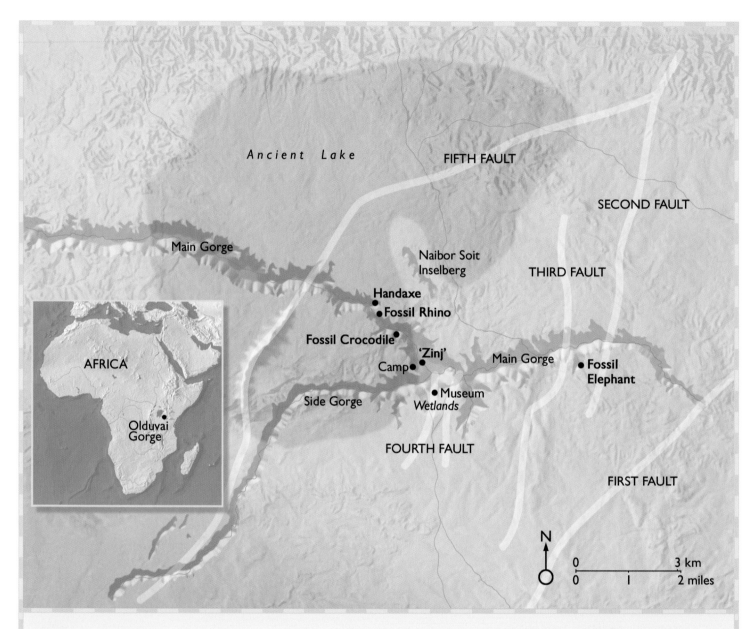

Ancient Lake

FIFTH FAULT

SECOND FAULT

Main Gorge

Naibor Soit
Inselberg

THIRD FAULT

Handaxe
• Fossil Rhino

Fossil Crocodile•

'Zinj'
Camp•   •

Main Gorge   • Fossil
Elephant

Side Gorge   • Museum
Wetlands

FOURTH FAULT

FIRST FAULT

AFRICA

Olduvai
Gorge

N

0   3 km
0   1   2 miles

Olduvai Gorge

One of the most famous anthropological sites in the world, Olduvai Gorge in Tanzania was, around two million years ago, a lake fed by several rivers. Situated within the Eastern Rift Valley, it was close to several active volcanoes that often showered volcanic ash over the region. The waters of the lake and the rivers that fed it attracted the plentiful game of the region and our ancient human relatives. They visited its shores looking for suitable materials to make their primitive stone tools and perhaps to scavenge carcasses for food.

The site was first explored by Louis Leakey in the 1930s but it was not until the 1950s that he

and his wife, Mary, set about the systematic excavation. The present river gorge has been cut down through the layers of Pleistocene (Ice Age) deposits by ephemeral flooding of the river. Now the deposits are exposed in the walls of the gorge from which they can be excavated and searched for fossils.

The Leakeys found the first Oldowan stone tools, lots of fossil animal bones but frustratingly few human related bones. It was not until 1959 that Mary Leakey found their first really important human related fossils – a fine but fragmentary skull. Luckily Mary was highly skilled at conserving and reconstructing such

broken remains. Louis gave the fossil a new genus and species name *Zinjanthropus boisei* or 'Zinj' for short (*Zinjanthropus* is derived from an Arabic name for East Africa and *anthropus* meaning 'man'). He was disappointed though, as with its heavy ape-like structure, it was not really what he hoped to find. He was looking for something that he really could call 'the missing link' between ancestral apes and ancestral humans. That turned up in the early 1960s but was even more fragmentary and it was several years before Louis was able to claim that he had indeed found what he was looking for and he called it *Homo habilis*.

Essentially, the difficulty arises from the fact that we are dealing with extinct fossil species from which only a limited number of mainly physical characteristics can be deduced. But even if we had the sort of quality and quantity of information about these species that can be provided by desiccated or frozen mummies such as Ötzi, the Neolithic 'iceman', I suspect that we would still be arguing the toss. Even if we were able to recover their DNA there might still be a dispute because species vary through space and time and some evolve from one into another. Apart from anything else, the question of what is, and what is not, a basic human attribute is a very emotional one – one which philosophers, theologians and scientists have argued about over the centuries.

The search for fossil remains of those ancient relatives who are on the borderline between an ape-like and a human-like condition was derailed several times after Darwin's death in 1882. As we have seen, there was Haeckel's claim that the ancestry of

The Piltdown Man hoax

In 1912 the discovery of a fossil jaw and part of a skull roof at Piltdown in Sussex, England was hailed by most of the experts of the day as the 'missing link' between apes and humans. Named *Eoanthropus dawsoni*, meaning Dawson's dawn man, it seemed to combine a large brain with a primitive ape-like jaw, and it was seen to vindicate the brain-led theory of human evolution and a great antiquity for humans.

Not until 1953 was it finally proven that the fossil remains were in fact a fairly crude forgery made from part of a modern human skull and the jaw of an orangutan. It is still not entirely clear who perpetrated the fraud; the finger has been pointed at a variety of people including Arthur Conan Doyle (author of the Sherlock Holmes detective stories), the Jesuit priest Teilhard du Chardin and Martin Hinton who was a colleague of Sir Arthur Smith Woodward at the Natural History Museum in London and held a grudge against him. But following a recent book showing that his entire life involved forgeries, it would appear that the culprit was Charles Dawson, one of the discoverers of Piltdown man.

humans was to be found in southeastern Asia, a supposition which seemed to be supported by Dubois' discovery of Java Man in the 1890s, followed by Peking Man in the 1920s. There was also the whole sorry saga of the Piltdown fraud, which did so much damage to the reputation of the whole science.

Not until the 1950s, and the Leakeys' African endeavours, was the search put back on a more scientific footing – although at times it must have seemed incredibly frustrating to Louis (1903-1972) and Mary Leakey (1913-1996). Louis had found masses of primitive stone tools at Olduvai back in the 1930s but it was not until 1959 that Mary and he found their first good human-related fossils. Although Louis initially thought that these beings (*Zinjanthropus boisei*, later placed in the genus *Paranthropus*) had made the stone tools, they were too ape-like to be his putative 'missing link' and so the search went on.

Brains define humankind?

One of the main criteria thought to be diagnostic for humankind was brain size, but the big question was: how big? Higher apes of today have brains that are only about a third of the size and weight of the human brain – around 400cc for chimpanzees and 500cc for gorillas compared with about 1350cc for humans.

However, human brain size varies considerably between about 900 and 1,700cc and was the subject

Olduvai Gorge in Tanzania was an excellent initial choice for excavation because the strata are clearly laid out with what geologists call a 'layer-cake' stratigraphy. The flat layers of ancient sediment lie undisturbed by complex earth movements and the erosion of the gorge has cut down them exposing long sections which can easily be investigated in detail. The sedimentary strata are interlayered with volcanic lavas that can be dated radiometrically so that we now know the oldest strata, that contain stone tools, date back some 1.8 million years. Many animal fossils have been recovered but frustratingly for the Leakeys, human related remains are remarkably rare here.

Brain size partly correlates with body size. When the brain sizes of a selection of our ancient relatives are compared we can see how it has increased over time in the human family but not in the higher apes. Most *Homo* species have brains greater than 650 cc in volume but little *Homo floresiensis* is down with the chimps and our most ancient ancestors.

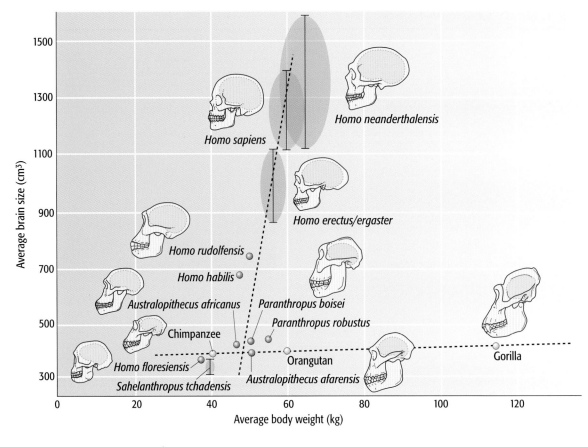

Arthur Keith's 'cerebral rubicon' became widely accepted.

of much discussion in the 19th century when it was thought that there was a simple correlation between brain size and intelligence. The fact that the average female brain size is slightly less than that of males was even used as apparent 'scientific' support for denying women the vote.

Eventually it was realized that average brain size within particular species correlates with body mass and since women have a lower average mass than men the difference is easily explained, and when corrections are made for mass there is no difference in brain size between the sexes. Neither is there any link between brain size and intelligence within a species.

Dart's australopithecines have a brain capacity that, at around 450cc, is not very much larger than that of a chimpanzee. By contrast the first extinct human relative to be discovered was the Neanderthal people (*Homo neanderthalensis*) and measures of their brain capacity showed little or no difference with modern humans; if anything their brains were slightly larger. Even *Homo erectus* has a brain capacity that Dubois measured at 850cc and is now known to range from 700-1,300cc. So what is the dividing line between ape-like and human-like brain capacity?

In the late 1940s, the British anthropologist Arthur Keith (1866-1955) argued that the line should be drawn at 750cc — midway between the largest known gorilla brain and the smallest human brain and this 'cerebral rubicon' became widely accepted.

Redefining the ape-human boundary

It was not long after the discovery of Zinj, or Nutcracker Man, that more human-related fossils turned up at Olduvai. From 1960 various hand and foot bones, a lower jaw and part of a skull roof were found. To Louis this was more like what he had been seeking. The teeth in the jaw were **clearly smaller** than those of Zinj but still more australopithecine than human-like. And what about the critical brain size? To his frustration Louis did not have a complete skull — only bits of the skull roof. Were these enough to calculate the original brain size and if so, would it be on the right side of Keith's 'cerebral Rubicon'? Louis had to rely on colleagues who were specialists in different aspects of anatomy to help analyse his finds. The foot and hand bones went to Peter Davis and Michael Day while the tricky task of reconstructing the skull roof and estimating the brain size went to Phillip Tobias and

John Napier. The latter two really had a problem in trying to determine the curvature of the brain case from just a few pieces and they knew what was at stake. They also knew that their academic colleagues would need convincing data and arguments if their estimate were to stand any chance of being accepted.

So they were not to be hurried, but Louis was impatient for a result and one that would satisfy him. Not until 1963 did the provisional result emerge – and at between 600 and 700cc it was just on the wrong side of the 'fence' for Louis. Then, when the final measure of 675-680 cc was given, it confirmed that the fossil could not be squeezed into the definition of the genus *Homo* as it then stood. However, the post-cranial bones seemed to be pointing in a more human direction and analysis of the hand seemed to indicate that this ancient relative had a fully opposable thumb – unlike the living chimpanzees – and could therefore have manufactured primitive stone tools. Less surprisingly, the leg bones indicated an upright standing biped with a walking gait, which although not fully modern was pretty close.

By 1964, Louis and his colleagues took the plunge and named the new find *Homo habilis*, a name suggested by Raymond Dart and meaning 'fit' or 'apt' man, though often translated as 'handyman'. This was a very bold move. Not only did it virtually triple the then known temporal range of our genus back to nearly two million years, but in doing so the team redefined *Homo* in favour of Louis' new find. Keith's baseline of 750cc was lowered to 600cc, thus helping to justify the placing of 'handyman' in the genus *Homo* instead of *Australopithecus*. In addition it was argued that 'handyman' had the ability to make use of tools and that such manufacture was a unique attribute of members of the genus *Homo*.

Not surprisingly there was considerable dissent among the experts over Louis' somewhat cavalier treatment of the genus *Homo* and indeed even today there is a significant number of palaeoanthropologists who prefer the designation *Australopithecus habilis*. At the time one of the lines of argument was that the fossil remains did not necessarily belong to the same individual, were far too few and too fragmentary to use in defining such an important baseline. Since then, however, more

fossils have been found that have been assigned to 'handyman', including a crushed skull found at Olduvai in 1968; it provided a better but even lower measure of its brain capacity at around 612cc – precariously close even to Leakey's baseline.

Tools 'maketh man'?

As we have seen, Louis Leakey was particularly interested in the possible role which tool-making might have played in the transformation of our ancestors and relatives from more ape-like forms into more human ones. The discovery within the lowest strata at Olduvai of primitive cobble tools, which Louis called the Oldowan technology, was perhaps his most important early find. Initially, there was no reliable radiometric date for these tools but later on a layer of basalt at the base of the Olduvai section was dated at around 1.8 million years and importantly it was found to record a reversal in the Earth's magnetic field (see box on page 120).

The first fossil remains of *Homo habilis* found by the Leakeys in the early 1960s were very fragmentary but Louis Leakey and his colleagues persisted in naming them as a new species and the oldest known member of the genus *Homo*. But this was only achieved by redefining the definition of *Homo*, much to the annoyance of some expert anthropologists.

Leakey experimented and showed how the tools might have been made.

The 1.8 million year old stone tools found in the lowest layers at Olduvai are very simple and are referred to as the Oldowan technology. Louis Leakey named his new species *Homo habilis* (meaning 'handyman') because he considered that they were the first manufacturers of stone tools.

At first Louis Leakey thought that his new robust australopithecine species *A. boisei* must have made the Oldowan tools but he changed his mind and considered the possibility that tool-making might have been a unique human attribute and connected to an increase in brain size. As we have seen, his description and naming of his subsequent find of a new human relative with an enlarged brain and more human features linked tool-making and a redefinition of the genus. Thus *H. habilis* became the Oldowan tool-maker.

Leakey's interest in tool-making included some innovative experimentation on how such tools could be made. In addition he realized that very little was known about the behaviour and abilities of our nearest living relatives, the chimpanzees, and that it was about time the situation was changed. Observation of chimpanzees in captivity was very limited in its usefulness, however, and here he was in Africa in the natural home of the chimpanzee. So, why not try and make field observations of them? He had no time to do this himself and so encouraged a young student to see what he could do.

There was a problem... Forest-dwelling chimpanzees can be remarkably elusive as they can move quickly over considerable distances, and they are very aware of intruders and simply absent themselves. So it's probably understandable that Louis' first observer tasked with obtaining data on chimpanzees gave up after failing to get anywhere near them to make any useful observations. However,

As Darwin noted, the first westerners to notice that chimps are capable of making tools were a couple of American missionaries to West Africa in the 1840s. Their observations were forgotten about until recently after this chimp behaviour had been rediscovered by Jane Goodall in the late 1960s. Goodall was a protégé of Louis Leakey's and pioneered the scientific observation of higher apes in the field. We now know that young chimps learn tool use from their mothers and so it is a cultural attribute, something that was thought to be a purely human acquisition.

Louis realized that young male researchers are often too impatient to get results whereas young women are often not only more patient, but also more sensitive to nuances of behaviour. So he advertised, and his new advertisement in the London *Times* led to the arrival in Kenya of a young Englishwoman by the name of Jane Goodall.

The rest, as they say, is history. Her work in the Gombe forest in northern Tanzania revolutionized the study of chimpanzee behaviour. Louis had been right, her patience and sensitivity paid dividends in the long run and one of her major discoveries, in the late 1960s, was that chimpanzees *do* use tools for a variety of purposes.

Strictly speaking, Goodall *re*discovered chimpanzee tool use as the chimpanzee use of stones to crack nuts had actually been reported by Darwin in *The Descent of Man* (1871) following even earlier observations by a couple of American missionaries to West Africa in the early 1840s.

We now know that tool-making is a cultural acquisition by chimpanzees which is taught generation to generation by mothers to offspring. Different chimpanzee populations use different tools, but new tool use can be introduced. The tool use can also be quite sophisticated and involve the careful selection of materials, their preparation or modification and even a two-stage process. Chimpanzee infants need considerable practice over considerable time (as much as five or six years) to become adept at using the more complicated tools such as an anvil and hammer to crack open hard nuts. But the nutritional dividends make it worthwhile. Their strong general tendency to literally 'ape' their mothers helps enormously and it may be that they are 'hard wired' for such behaviour.

The fact that chimpanzees can make and use basic tools means that tool use cannot be ascribed to us as a uniquely human attribute. It is quite likely that the common ancestor of the chimpanzees and humans was also capable of such a basic tool culture around seven million years ago. The problem is that such basic tool use is unlikely to fossilize because most of the 'tools' are organic such as twigs (for fishing termites out of their nests) or leaves (as water collecting 'sponges') which do not preserve as fossils in the rock record or, if they are preserved, cannot be recognized as 'tools'. Even the more

durable wood and stone tools for nut cracking are not modified in any really significant way. However, some observers of the Taï Forest chimpanzees in the Ivory Coast have claimed that repeated use of certain nut-cracking sites can leave a potential archaeological signature of such behaviour which might possibly be picked up in the fossil record. But whether or not this is so remains to be seen.

The revelation of chimpanzee tool-making activities cast doubt on the likelihood that the manufacture of stone tools might be a unique human aptitude. By the 1970s the record of stone tool manufacture was being extended through the newly discovered sites of the late Pliocene and early Quaternary age in Africa, especially in Ethiopia. In 1973 a German archaeologist Gudrun Corvinus, working with the Franco-American expedition which discovered 'Lucy', found and documented dozens of localities with abundant stone tools exposed on the surface of the landscape in Ethiopia's Hadar region and Gona River area. But with all the political unrest and subsequent civil war in the country, it was not until the early 1990s that fieldwork resumed.

The oldest stone tools

A mainly American team led by Ethiopian archaeologist Sileshi Semaw, now at Rutgers University, made a careful study of the Gona River area where stone tools occurred *within* the stratigraphic sequence of ancient sediments and could therefore be placed within the known chronology with some degree of accuracy. In contrast, any tools that are found lying on the surface cannot safely be placed in such a context. In addition, geologists could analyze the sediments in which the tools were embedded and then reconstruct the original environment. In this case the original environment turned out to consist of river floodplains near river channels that contained a variety of rocks. From these the tool-makers had a good choice of different rock materials, being able to select those that were hard enough to retain an edge for a reasonable time, and could be knapped in a satisfactory manner.

Within a few square metres at Gona some 2,970 stone tools were found and mapped out, along with flake debris and the original rock cores from which the tools had been made. Also found was a battered

hammer stone which had been used to manufacture the tools. Over half of them were made from a glassy kind of lava called trachyte. It is hard, brittle, and flakes can be struck from it to leave sharp edges suitable for cutting even tough materials. The tool manufacturers knew their rocks and had carefully selected the best material.

The basic style of stone tool is generally similar to that found by Louis Leakey at Olduvai and which he had called the Oldowan industry, but there are no retouched flakes at Gona so in one sense the latter tools are slightly less diverse and therefore perhaps more primitive. Fortunately, adjacent volcanic ash layers allowed the radiometric dating of the sediments at between 2.9 and 2.5 million years – which is several hundred thousand years older than the Olduvai tools. These are still the oldest known stone tools in the world but as they show a degree of sophistication in their manufacture, there must have been a considerable previous history of experimentation and development by the beings that made them in the first place. The big question is who made them?

Making a stone tool that would do the job began with selecting the most suitable material.

Gona stone tools – The dating of some primitive stone tools from Gona in Ethiopia at between 2.6 and 2.5 million years old showed them to be the oldest known. It is not clear who made them but it may have been either an australopithecine or an early *Homo* species.

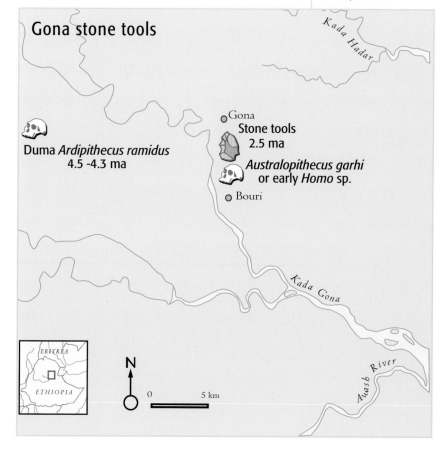

Gona stone tools

Kada Hadar

Gona
Stone tools
2.5 ma

Duma *Ardipithecus ramidus*
4.5 -4.3 ma

Australopithecus garhi
or early *Homo* sp.

Bouri

Kada Gona

ERITREA

ETHIOPIA

N

0   5 km

Awash River

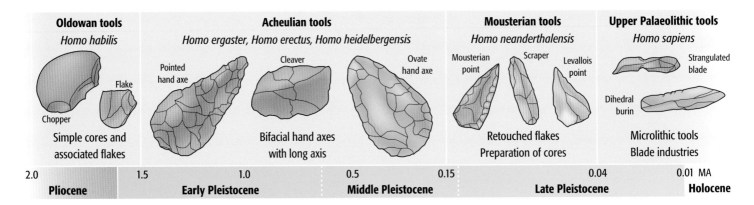

Oldowan tools	Acheulian tools	Mousterian tools	Upper Palaeolithic tools
Homo habilis	*Homo ergaster, Homo erectus, Homo heidelbergensis*	*Homo neanderthalensis*	*Homo sapiens*
Chopper, Flake	Pointed hand axe, Cleaver, Ovate hand axe	Mousterian point, Scraper, Levallois point	Strangulated blade, Dihedral burin
Simple cores and associated flakes	Bifacial hand axes with long axis	Retouched flakes Preparation of cores	Microlithic tools Blade industries

2.0	1.5	1.0	0.5	0.15	0.04	0.01 MA
Pliocene	**Early Pleistocene**		**Middle Pleistocene**		**Late Pleistocene**	**Holocene**

Stone tool types can be broadly arranged in a general 'evolutionary' sequence from the primitive Oldowan culture through to more advanced Upper Palaeolithic types. However, it is not possible to say who exactly made the different types until *Homo sapiens* was the sole surviving species.

A young Richard Leakey posing with one of the first (*Paranthropus boisei*) of the many spectacular fossil finds he was to make, especially in northern Kenya and Ethiopia.

Was stone tool technology 're-invented' twice by totally independent groups?

The first tool-makers?

Unfortunately the manufacturers of the tools did not leave any form of archaeological 'calling card' that might allow us to identify who exactly they were. The riverside Gona site was primarily one where they selected stones and made the tools but not one that they occupied for any length of time. However, nearby at Hadar some human-related bones have been found in association with Oldowan-style tools. The human relatives have been identified as australopithecine but in 1994 some 2.3-million-year-old stone tools and an upper jaw fragment were found – and the jawbone looks distinctly more human-like.

There is no doubt that both australopithecines and transitional more human-like species such as *A./H. habilis* and *A./H. rudolfensis* coexisted around two million years ago but there is no definite proof as to who made the tools. However, there is even another possibility, which might at first seem unlikely...

The Oldowan stone technology remained very conservative from its inception over 2.6 million years ago for well over a million years and the only human relatives who were equally conservative over this time range were robust australopithecines – members of *Paranthropus boisei* who lived in the region. If it turns out that they were the tool-makers, it certainly would be a 'turn up for the books' since this group are normally regarded as an evolutionary cul-de-sac with rather limited cognitive abilities – a bit like gorillas in comparison with chimpanzees. Furthermore, it would mean that stone tool technology had to be 'invented' twice by totally independent groups – the paranthropines and then again by more human-like relatives. It is another intriguing and unresolved puzzle.

The human from Lake Rudolf – KNM-ER 1470

Richard Leakey, Louis' son, had quickly established himself and his field team as very successful 'bone hunters', especially at Koobi Fora on the eastern shore of Lake Rudolf (now Lake Turkana) where there were rich pickings in late Pliocene to early Quaternary sedimentary deposits. Since these strata were of similar age to those at the base of the Olduvai section there was a chance that better specimens of 'handyman' would turn up at Koobi Fora.

In the three years from 1969, Richard and his team had recovered the partial remains of over 35 fossil individuals, more than his parents had recovered in 30 years of searching. And, in 1972, the year Louis died, Bernard Ngeneo, one of Richard's Kenyan field assistants, found a skull-shaped heap of broken bones, which were given the identifying tag

KNM-ER 1470 (KNM refers to Kenya National Museum and ER to East Rudolf where it was found). Alan Walker (1938-), a British anatomist and Meave Leakey (1942-), Richard's British wife took on the arduous task of reassembling the very complicated 3D jigsaw puzzle.

The face turned out to be quite wide and a bit australopithecine-like, but the dome of the skull was more inflated and human-like. Furthermore, there was no bony crest to the top of the skull, nor prominent brow-ridges as seen in the australopithecines. However, the teeth seem to have been quite large as in the australopithecines so there is a confusing mixture of primitive and more advanced features. The really tricky bit of the reconstruction was that of the face as it was hard to determine the forward slope of the facial bones. A low angle would produce an australopithecine-like forward projection while a higher near vertical angle would produce a short, more human-like appearance.

Initial field estimates of the brain size came out at around 800cc, well onto the human side of the divide even by Keith's standard, though more accurate calculations brought it down to around 750cc and therefore more marginal. Luckily, Louis lived just long enough to hear what his son had found. By this time the old tendency to name as a new species any new find with the slightest difference to what had been found before was frowned upon so Richard and his team held their fire. One of the problems was that the initial dating of a volcanic ash lying adjacent to the fossil layer produced an age of around 2.6 million years. This made it older than *Homo habilis*, which was dated at around two million years old. How could a larger-brained and therefore apparently more advanced human relative appear before a smaller-brained one? It did not seem to make evolutionary sense.

However, there was a problem with the dating and the disagreement turned into a rather acrimonious row which took some years to resolve with the date being revised upwards and closer to two million years old. It was not until 1989 that the skull was graced with a new species name – *Homo rudolfensis*. Importantly, it showed that there were other ancient human relatives, more advanced than the australopithecines but not as advanced as the Asian human relatives (Java and Peking Man,

technically known as *Homo erectus*) who had lived in Africa around the beginning of Quaternary times.

Nariokotome Boy – KNM-WT 15000

In 1984 Kamoya Kimeu, one of Richard and Meave Leakey's team of highly skilled fossil finders, made one of the great discoveries in the history of the search for our ancient human relatives. At Nariokotome on the western shore of Lake Turkana (formerly Lake Rudolf) he found the most complete ancient skeleton known – even better than 'Lucy'. Not only is most of the post-cranial skeleton still preserved but there is also an excellent skull including the lower jaw and most of the teeth. All that is missing are the feet and hands, although some finger bones were recovered.

As so often happens, experts disagreed – this time rather acrimoniously over the dating of the remains.

Reconstruction of the 'Nariokotome' boy's skeleton, one of the best preserved of an ancient human relative some 1.9 million years old.

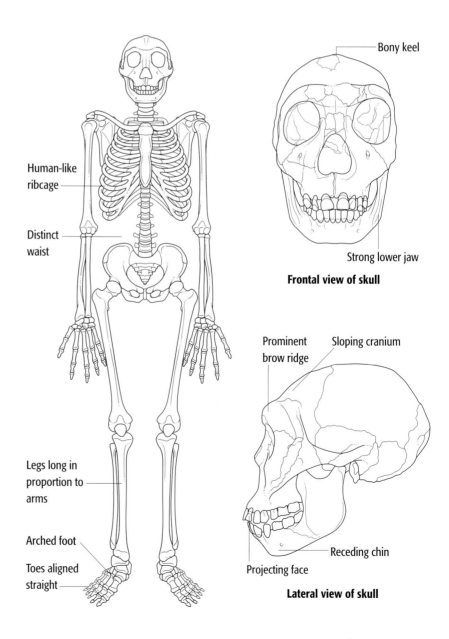

Human-like ribcage

Distinct waist

Legs long in proportion to arms

Arched foot

Toes aligned straight

Bony keel

Strong lower jaw

Frontal view of skull

Prominent brow ridge

Sloping cranium

Projecting face

Receding chin

Lateral view of skull

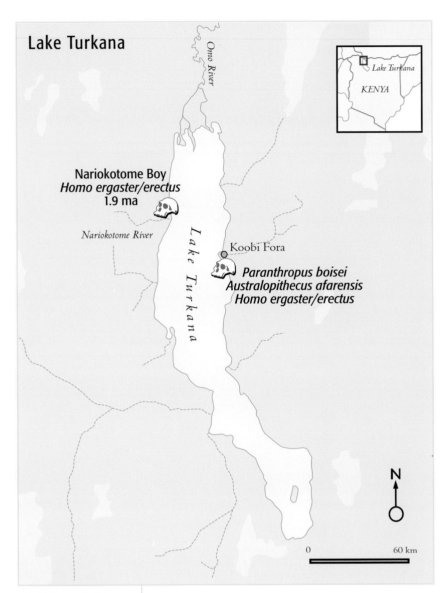

Lake Turkana – A whole hoard of scattered fossils have been found in Pliocene age deposits between 1.8 and 5 million years old around Lake Turkana. Apart from numerous animal fossils, the remains of three different human related species have been found.

Nariokotome Boy has a fascinating mix of features and attributes, some advanced, some primitive.

When reconstructed the skeleton stands about 160cm (5 ft 3 ins), a fair bit taller than any of the australopithecines of whom the tallest known is a male *A. afarensis* at about 151cm. But most surprisingly, examination of the Nariokotome Boy's teeth show that he probably grew more rapidly than modern humans. Although direct comparison of his development with modern humans would suggest that he was about 11 years old when he died, analysis of growth lines on his teeth suggest that he may only have been eight years old when he died. Had he lived to maturity, he would probably have reached a height of 185cm (6 ft 1in) and weighed about 66kg (145 lbs). But his brain capacity was only 826cc, perhaps rising to around 871cc on maturity. His body proportions are like those of modern humans who still live in northern Kenya and

Ethiopia – tall, long-limbed and slender-hipped – good distance runners today and perhaps in the past.

The big difference between Nariokotome Boy (alternatively known as Turkana Boy) and modern humans is in his skull. The face is quite human-like but retains some primitive features. It seems relatively large because he has quite prominent bony brow-ridges behind which the top of the cranium slopes back so that there is no forehead. The lower part of the face projects forward to the mouth and then recedes back into the lower jaw, which is strongly built. The muscularity of his jaw mechanism is emphasized by the remnants of a primitive bony keel on top of the skull for attachment of the temporal muscle sheets as seen in many of the australopithecines. The skull floor shows some buckling that could be related to a lowering of the larynx and the production of complex sounds and perhaps vocalization that perhaps included basic speech type sounds. Although his brain size is relatively small it is still bigger than any australopithecine brain. Importantly the structure of the inner ear that has to do with hearing and balance is human-like. It seems that vocalization and hearing were significant adaptations for the Nariokotome Boy and were perhaps connected to socialization and hunting.

Preservation of his ribs allows the reconstruction of his rib cage and chest region. Unlike the barrel-chested australopiths with their flared ribcages, the Nariokotome Boy had a very human-like chest, which was flattened at the back and front. And, there are indications that he had a distinct waist, suggesting a small stomach and gut from a change in diet compared with his plant-eating ancestors. He must have been eating a protein-rich diet and his body form supports the idea that his fellow beings were active hunters.

The structure of his backbone provides further support for the idea. Individual vertebrae have wear surfaces indicative of a habitual vertical stance and prolonged walking or running. Interestingly, the opening in the individual vertebrae for the passage of the central nerve cord is unusually small compared with that of modern humans. Our modern nerve cord is enlarged because the spinal nerve bundle supplies not only the legs and lower part of the body but also the chest and diaphragm. The latter not only control breathing and synchronize it during exercise but also allow us to talk while walking or running so there are a lot of nerves to coordinate all the activities. The fact that the Nariokotome Boy had a significantly smaller bundle of spinal nerves suggests that he was not greatly engaged in one of these activities. Speech is the most likely one to be missing and it is likely that his vocalization was quite limited compared with modern humans – but it still represents a significant advance on that found in the higher apes today.

Altogether the Nariokotome Boy has a fascinating mixture of primitive and more advanced features and attributes. As Alan Walker, the chief investigator of the skeleton has pointed out, from a distance he would have looked just like a modern human moving through a landscape but Walker imagines that on meeting his gaze he would be met with that 'deadly unknowing I have seen in a lion's blank yellow eyes...he was not one of us'.

Anatomically, he is indeed human and perhaps one of our ancestors, but cognitively he is likely to have had very little consciousness. A modern human would have been little more than a curiosity and potential food object for him.

When it came to giving Nariokotome Boy a scientific name to place him in the chronological, evolutionary and classificatory pattern of our ancestors and relatives there has been considerable disagreement among experts. Some place him in the species *Homo ergaster*, which was first described in 1975 for African fossils which previously had been placed in *Homo erectus* (see page 108), a species defined originally from Asia.

As we have seen, experts disagree on the validity of the species *Homo ergaster*. Much of the problem arises from the question of whether it is likely that around two million years ago a population of human relatives dispersed across Asia from Georgia to China and Java and remained part of the same interbreeding species as another population in Africa, even if they did originate from the latter. Given the huge geographical distribution of relatively small populations with temporal spread from around two million years ago to around 100,000 years ago, it seems unlikely. And if it is unlikely then it may be useful to distinguish the African members as a separate species *Homo ergaster*,

providing there is some discernible anatomical justification for the separation, which can be distinguished in the fossil bones.

Other experts argue that this is the problem: the variation found in known members of *Homo erectus* from Asia includes the slight differences to be seen in the African specimens and that means therefore they should all be included in *Homo erectus*. The picture has been further complicated by recent finds at Dmanisi in Georgia.

In summary, there is good evidence from Africa for the progression from more ape-like australopithecine relatives to more human-like ones even if it is, as yet, fragmentary and not as straightforward as we might like it to be. It turns out that the diversity of bipedal but small-brained australopithecines, which occupied a broad swathe of Africa from three to around 1.5 million years ago were also contemporary with at least two transitional species *Australopithecus/Homo habilis* – *rudolfensis* with enlarged brains around two million years ago and, a third more convincing member of our genus *Homo ergaster/erectus*.

Certainly, the latter can be seen in retrospect to have by far the greatest potential as an ancestral form to successive human ancestors but if so, from where did *Homo ergaster/erectus* arise? The two main options are either *A./H. habilis* or *A./H. rudolfensis* and, opinion is divided over the issue. For some time now the only obvious potential ancestor was *Australopithecus afarensis* but with the discovery of older relatives such as *Ardipithecus*, *Orrorin* and *Sahelanthropus* there are some alternatives. At the moment the most serious contender is *Ardipithecus*.

As we have seen, the old ideas about the essentials of humanness have been swept aside in recent decades. Bipedalism was rejected a long time ago. Brain enlargement, however we mark the watershed, may be part of the story, but it cannot be *all* of the story because we have clear anatomical members of *Homo*, such as *H. ergaster/erectus*, which still have relatively small brains. However, brain enlargement and reorganization connected to speech and increased socialization does become an essential ingredient later on. As with almost all major evolutionary innovations, the interesting question is what benefit did the initial adaptation bring and what 'fuelled' it?

KNM-ER 1470, a famously problematic skull, found by Richard Leakey, which lies close to the divide between the genus *Australopithecus* and *Homo*, and *Homo* has been named as *Homo rudolfensis*.

CHAPTER SIX

The southern apes, the australopithecines

 Today some seven or eight different kinds (species) of australopithecines, meaning 'southern ape', are recognized. All have been discovered over the last 80 years in Africa and they include some of the best known of our ancient extinct relatives such as Australopithecus afarensis, better known by 'her' nickname 'Lucy' and 'Zinj' also known as 'Nutcracker man' or 'Dear Boy' but taxonomically known as *Paranthropus boisei*.

Excavation of fossils at East Turkana in northern Kenya. The site is rich in hominid and other fossils. The East Turkana expedition began in 1969 and was led by Richard Leakey.

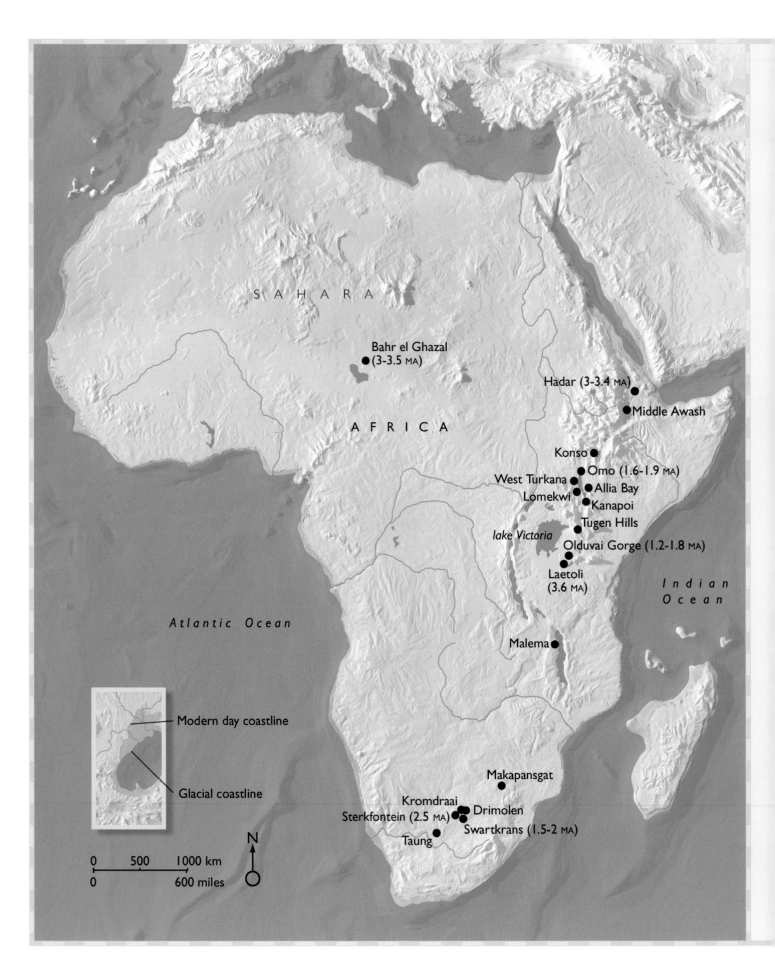

Bahr el Ghazal
(3-3.5 MA)

Hadar (3-3.4 MA)

Middle Awash

Konso

Omo (1.6-1.9 MA)

West Turkana
Lomekwi

Allia Bay

Kanapoi

Tugen Hills

lake Victoria

Olduvai Gorge (1.2-1.8 MA)

Laetoli
(3.6 MA)

Malema

Makapansgat

Kromdraai

Drimolen

Sterkfontein (2.5 MA)

Swartkrans (1.5-2 MA)

Taung

SAHARA

AFRICA

Atlantic Ocean

Indian
Ocean

Modern day coastline

Glacial coastline

N

0   500   1000 km
0   600 miles

Southern apes of Africa – *Australopithecines*

With more than six species recognised so far, the australopithecines were the most diverse and numerous of our extinct human relatives. Originally named by Raymond Dart in South Africa for his little Taung skull, the genus *Australopithecus* means 'southern ape'. And although it was many years before his genus was generally accepted, by the 1970s it seemed to becoming a pivotal genus in the early evolution of the human lineage, the only question seemed to concern which of the australopithecines gave rise to the genus *Homo*?

For over two million years these small bipedal apes spread through the woodlands and forests of eastern Africa and as far west as Chad. Some of them had powerfully muscled jaws for eating tough plant materials and became known as the 'robusts' and were even given a separate genus name *Paranthropus*.

In contrast, the others, called the 'graciles', had, as the nickname suggests, less heavily built skulls and evidently ate a more mixed diet that probably included some flesh. Generally, the ranges of the robusts and graciles overlapped except in the most northeasterly region of Ethiopia where they seem to have been all 'graciles'.

Dart had always claimed that his *Australopithecus africanus* had been capable of walking upright and it was subsequently proved that he was right. Despite their small size of around a metre, and small ape-like brains, these ancient relatives definitely were capable of walking although some of them also retained an agility for climbing. Some of them may even have been capable of making stone tools but we do not really know.

The australopithecines span an interval of three million years of late Pliocene to early Pleistocene times. Within this span there were times such as 2.5 million years ago when there were at least three contemporary species of australopithecines living in eastern Africa and they overlapped with another one or two species of our own genus, *Homo*. Altogether they present a radical new picture of the pattern of human evolution. No longer can we hold the old-fashioned notion of a simple linear progression of advancement from a more primitive to a more advanced state. Our family 'tree' is in fact a shrubby bush within which the evolutionary connections are still very far from clear. Some experts even claim that our human evolution might have bypassed the australopithecines altogether.

The Taung 'child'

The first australopithecine to be found was *A. africanus* and was described in 1925 by Australian anatomist, Raymond Dart (1893-1988), who was domiciled in South Africa. Dart had been an army medical officer and a veteran of WWI. Following demobilisation he was employed as a research assistant to fellow Australian academic anatomist

Raymond Dart described a new genus and its first species - *Australopithecus africanus* from the juvenile Taung skull portrayed here. It was many years before this important new genus was accepted and by then Dart and his ardent supporter Robert Broom had discovered many new specimens from the South African caves around Sterkfontein and Swartkrans.

Grafton Elliot Smith (1871-1931) in London. Encouraged by Smith, Dart soon took up a post as professor of anatomy in the University of Witswatersrand in South Africa. Due to the lack of teaching material he encouraged his students to keep an eye and ear open for any primate skeletal specimens either modern or fossil. Eventually in 1924 he was sent a box of limestone rock with bits of bone embedded in the rock. The specimens came from one of a number of working limestone quarries at Taung, southwest of Johannesburg.

There is a nice story attached to the arrival of the box at Dart's home. Apparently he was all dressed up and about to leave the house to attend the wedding of a friend of the family but could not resist the temptation to open the box. In one of the first specimens he was intrigued to see some well-preserved facial bones obscured by the surrounding rock and had to be dragged away by his wife. True or not, it is a typical 'folk' story of the dedicated scientist who will subjugate all other responsibilities for the sake of his scientific curiosity.

It took Dart many hours of painstaking work to remove the hard rock matrix from around the bone, again the story is that he used his wife's knitting needles, which he sharpened to form small chisel-like implements. He was kept going by the realization that it was a remarkable little face that was slowly emerging from the rock. Eventually, when he was finished he had a small but curiously human-like and complete face with lower jaw and teeth. In addition, the whole of the brain was represented by a rock endocast, which fitted in place behind the face. The extraordinary fact was that the whole block had survived being blasted from the original rock by the quarrying activity and had been handed to the quarry manager who, having heard of Dart's interest in such material, had then passed it on to him.

Thrilled to have such a prize specimen Dart wrote up his description and conclusions for the prestigious international science journal *Nature* in 1925. He made some far-reaching claims for the little immature skull. To begin with he described it as belonging to a whole new genus and species of extinct human relative which he called *Australopithecus africanus*, meaning 'southern ape from Africa'. To Dart the skull showed a combination of both ape-like and human features even though he also

recognized it was a juvenile retaining some milk teeth along with newly erupted permanent teeth. Indeed he even specified a possible age of about seven years old for the 'child' based on a human model of dental development. Modern research however shows that when a more appropriate australopithecine model of tooth eruption is applied, the Taung infant was only about three years old when it died. The canine 'eye teeth' are relatively small and human-like.

The facial aspect is quite vertical with relatively little muzzle projection (prognathism). There is a forehead and little or no bony brow-ridge. Dart thought that the brain shape was human-like and that the position of the foramen magnum (opening for the spinal nerve cord) at the base of the skull showed that the creature held its head up and walked upright. In addition he claimed that these small 'ape-men' were predatory carnivores who lived in caves and hunted game which they took back to their caves to consume. A major problem for Dart was the indeterminate age of the Taung find. In these pre-radiometric dating days, it was particularly difficult to establish the age of many cave deposits, especially ones that had been blasted out of their original stratigraphic context.

Unfortunately for Dart, despite his connections through Arthur Keith (1866-1955) with the palaeoanthropological community back in London, his dramatic claims were a 'step too far' to gain acceptance. Coming from someone who was a newcomer, and, moreover, one based in a remote outpost of the European and American dominated scientific world, they were simply too revolutionary. As Darwin knew only too well, to gain any acceptance of a radical new idea or discovery, it was necessary to prepare the intellectual ground within a network of influential and trustworthy colleagues who would provide enough support to give the new idea any chance at all of survival. Dart had hoped that his old mentor Smith and professor Keith would provide such support but they did not.

Critics dismissed the skull as merely that of a juvenile chimpanzee and there were some grounds for this criticism. The skulls of immature higher apes, even gorillas, do look much more human like than those of the adults with all their secondary sexual characteristics and sexual dimorphism.

Dart's claims were 'a step too far' and getting them accepted by the scientific community proved to be a challenge.

Features such as bony brow-ridges, crest ridges on the skull roof and facial prognathism (projecting lower jaw) become much more prominent as the apes become adolescent and pass into maturity. In males the features are even more exaggerated.

Dismayed by the negative reaction to his 'Taung child', Dart thought that the only answer was to take his specimen to London and show it to Keith and anyone else who mattered. But Dart had not realized the extent to which the British and other European experts had become 'hung-up' on Eurocentric ideas for the origin of humanity. Although the now notorious and discredited 'Piltdown Man' had been found in 1912, it was still at this time very much in favour and Keith was one of Piltdown's main protagonists.

When he got to London in 1926, Dart had a stroke of bad luck. His arrival coincided with the arrival of 'Peking Man' on the international stage. As we have seen, Haeckel's theory of an Asian origin for mankind was not yet dead and indeed was considerably bolstered by the discovery of 'Peking Man' (known then as *Sinanthropus pekinensis*) in China by an international team led by the Canadian anatomist Davidson Black (1884-1934) and funded by the American Rockefeller Foundation. Although *Sinanthropus* was initially based on just a few teeth, it too was published in *Nature* and became headline news.

The only good news for Dart was that back in South Africa he found a doughty champion by the name of Robert Broom (1866-1951). An ex-pat Scotsman turned palaeontologist in his retirement, Broom had no time for the niceties or opinions of English academic experts and firmly believed in the authenticity of *A. africanus*. He was determined to find more specimens that would support Dart's evidence and ideas, and he targeted some other limestone caves and quarries around Sterkfontein, north of Johannesburg, as potential sources of good primate fossil material.

It took Broom some 10 years but in the end he certainly came up with the 'goods' – a wonderfully well preserved skull. This time it was clearly that of an adult but it appeared to be very different from Dart's little Taung specimen. The face was very broad with a flat forward projection and fairly heavy brow-ridges that sloped back into the roof of the

The photomontage shows Robert Broom (in dark waistcoat) with some of his sketch reconstructions, specimens and many published papers describing them, including the first of the robust australopithecines.

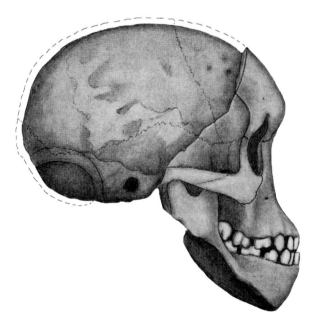

A side view of Dart's Taung 'child' shows the well preserved face, on which the bony brow-ridge has not yet developed, and the stony endocast of the brain cavity.

Cradle of Humankind

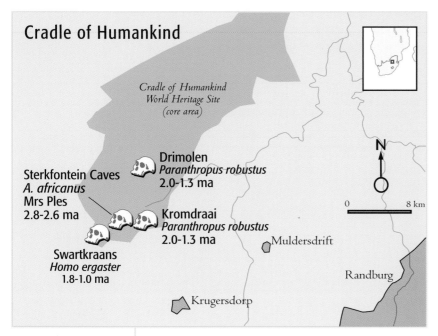

Cradle of Humankind
World Heritage Site
(core area)

Drimolen
Paranthropus robustus
2.0-1.3 ma

Sterkfontein Caves
A. africanus
Mrs Ples
2.8-2.6 ma

Kromdraai
Paranthropus robustus
2.0-1.3 ma

● Muldersdrift

Swartkraans
Homo ergaster
1.8-1.0 ma

Randburg

N

0   8 km

▲ Krugersdorp

Cradle of Humankind –
Several of the numerous cave sites in the Johannesburg and Pretoria region of South Africa that have yielded so many fossil finds of our human relatives have now been recognised as a World Heritage Site.

Mrs Ples is the popular name of the world's most complete skull of *Australopithecus africanus*, discovered by Dr Robert Broom and John Robinson in April 1947. Only a short time afterwards, a partial skeleton was discovered very close to where the skull had been found. This specimen is known by its catalogue number, Sts 14.

Was Dart correct in allowing his theories to be influenced by humankind's propensity for violence?

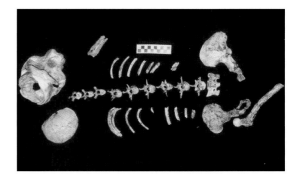

skull without any sign of a forehead and it had a small ape-sized brain. However, it also had distinctly un-ape-like teeth – large flat molars and small canines. And again the position of the foramen magnum on the base of the skull suggested an upright head position relative to the backbone and therefore bipedalism. The new find was duly given a new genus and species name – *Plesianthropus transvaalensis*, meaning 'near ape from Transvaal' – nicknamed 'Mrs Ples'. Using a gorilla model, the lack of a bony crest ridge to the skull indicated that it was a female as only the much bigger (sexually dimorphic) males present this feature. Today 'Mrs Ples' has been repositioned as an adult of Dart's original species *A. africanus* and Sterkfontein has been declared a World Heritage Site.

Broom went on to find more fossil-bearing limestone caves near Swartkrans and Kromdraai, all near Johannesburg and a spectacular series of jaws, teeth and the occasional skull were recovered. Many of the fossils were well preserved by the surrounding limestone which has a similar chemical composition to bone and teeth, indeed some of the teeth look almost unbelievably fresh and, apart from their size, could have come from a modern dentist's surgery. Another plus is the fact that the fossils are rarely deformed but retain their original shape. Broom thought that these fossils represented yet more new species and genera, and indeed some did look rather different from 'Mrs Ples' with even more massive

facial bones and features such as a prominent bony crest to the skull that has no forehead whatsoever and a very low dome.

Today, however, only two different forms are recognized: 'Mrs Ples' represents the so-called gracile (graceful) species *A. africanus* and the so-called robust form is *Paranthropus robustus*, although some experts still regard it as just another species of the genus *Australopithecus*.

By the late 1940s Broom and Dart found yet another cave site at Makapansgat, northeast of Johannesburg, where they found yet more fossils of their gracile form, but Dart noticed that at this site many of the bones were damaged. By this time he had lived through two world wars and like so many people of the time was still reeling from the overwhelming evidence of humankind's potential for destruction and genocide. Dart interpreted the damaged skulls and jawbones as signs of internecine aggression among the australopithecines who apparently used bones as weapons against one another – an activity for which he coined the rather pedantic academic phrase 'osteodontokeratic' (meaning bone-tooth-horn') culture.

Dart continued to develop these ideas in the latter part of his life. They were given an enormous boost when popularized by the American writer, dramatist and anthropologist Robert Ardrey (1908-1980) whose best-selling books *African Genesis* (1961) and *The Territorial Imperative* (1966) introduced the world to the 'killer ape' hypothesis and the notion that we humans are 'Cain's children'. Ardrey also used the growing information about animal social behaviour promoted by the German ethologist Konrad Lorenz (1903-1989) in popular books such as *King Solomon's Ring* and *On Aggression*.

However, today the material basis for Dart's claims has been reassessed and largely found wanting.

Much of the damage to the bones that does indeed look like impact fracture and breakage is actually post-mortem. Cave environments are commonly subject to roof falls of rock fragments, which is a more likely cause of the damage. But some of the skulls do show clear signs of animal predation including toothmarks, which could well have been made by a predatory big cat. Today leopards drag their prey by the neck or head up into trees beyond the reach of other competing predators such as lions or hyenas. It is highly likely that the small and relatively defenceless australopithecines would have constantly suffered from such predation.

One aspect of Dart's analysis that has stood the test of time is his claim that the australopithecines used broken bones as tools. Stick-shaped bones were excavated from Swartkrans cave between 1965-83 by his student C. K. (Bob) Brain and simple pebble tools have been recovered from both Sterkfontein and Swartkrans.

The authenticity of Dart's discoveries and the potential role of the australopithecines in the human story were greatly enhanced in 1947 when the eminent British palaeontologist Wilfrid Le Gros Clark (1895-1971) visited Johannesburg to see Dart's collections for himself. He made a detailed study of the remains, which by this time included bones and teeth from at least 30 different individuals. Le Gros Clark reported in *Nature* that Dart and Broom's conclusions 'had been entirely correct in all essential details'.

Such was Le Gros Clark's reputation within the academic anthropological network that Dart and his ideas were at last accepted along with the expectation that Africa had been the 'cradle of humankind' as Darwin had predicted. Le Gros Clark also joined forces with Joseph Weiner and Kenneth Oakley (1911-1981) to expose 'Piltdown Man' as a fraud in 1953 (see box on p. 125) which again helped reinforce the acceptance of Africa as the core site of human evolution and the position of the australopithecines as genuine human relatives.

The overall picture that the South African australopithecines presented was one which seemed to support the old idea of our ancestors as being 'cavemen' even in the early stages when they were still very much also 'apemen'. Although it was becoming accepted that they did walk upright, with their small

Even today big cats such as leopards frequently prey on apes and drag their carcasses up into trees to prevent scavenging by the likes of hyenas. Tooth and claw marks found on many fossil skulls show that our extinct human relatives were also often predated by big cats.

size and small brain they were also thought to be very ancient despite the problem presented by the determination of just how ancient they were. With their relatively small teeth they were thought to be meat eaters and probably hunters who killed game and brought it back to their caves to be consumed much as cave-dwelling hyenas did.

The general consensus that emerged during the 1920s was that although humans and the higher apes shared a common ancestry, the original divergence happened a long time ago perhaps as far back as Miocene times. In those pre-radiometric dating days that was thought to be at least several tens of millions of years ago. By the late 1940s the first attempt at a radiometric-based timescale by British geoscientist Arthur Holmes (1890-1965) indicated that Miocene times started about 32 million years ago. That still seemed to give plenty of time and evolutionary 'distance' between the primitive ape-like australopithecines and 'real' humans. The development that was to make the big difference between them and us was that of the brain. Therefore the real 'missing link' that was yet to be found was a 'brainy' or intelligent, ape-man who walked upright. The question was: what point in the development of brain size should be taken as the critical turning point between them and us? It is a question to which we shall return.

The real 'missing link' that was yet to be found was an intelligent bipedal ape-man.

Opening up our African ancestry

The story of the australopithecines in Africa took on a whole new aspect in the 1940s and '50s thanks to the activities of the energetic son of English missionaries in Kenya. Louis Leakey was brought up in Kenya, learned the local languages and customs and then went to the University of Cambridge where he was able to develop his enthusiasm for archaeology and his ambitions to search for the fossil remains of our ancestors back in Africa. He was seen as a young man to be encouraged and helped. From his background knowledge and reading Leakey knew that just before WWI German scientists had found human-like bones in the Olduvai Gorge regions of what had been German East Africa. In the post-World War I situation the territory was then the British Protectorate of Tanganyika (now Tanzania).

With the help of Hans Reck, one of the original German team, Leakey had, in the 1930s, relocated the approximate locality of the original find. The geological situation in Olduvai seemed very promising in the search for such fossils. The sides of the dry ravine revealed a succession of sedimentary layers lying flat one upon another in what geologists call a 'layer cake' succession with the youngest at the top of the pile and the oldest at the bottom. The sequence was numbered from Bed I up to Bed IV and numerous hand-sized pebbles were found in Bed I, some of which were broken to form sharp angular edges that can function as very crude chopper-type tools. Louis named this stone tool technology as that of the 'Oldowan' culture and for many years they were the oldest stone tools known in Africa.

Reck's earlier find was thought to have originated from Bed II but in these early days of field investigation there was not a full appreciation of the potential complexities of the ways in which human- and animal-related remains can be recruited into the sedimentary record. Louis in his initial enthusiasm did not see that Reck's find could have been much younger remains, which had been displaced into its apparently much older situation, but detailed investigation of the sedimentary context proved that this had indeed happened. Louis began to investigate other potential sites such as Kanjera and Kanam, near Lake Victoria, where the remains of many extinct animals were found and some more possible human-related ones.

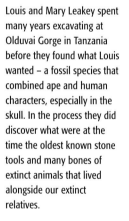

Louis and Mary Leakey spent many years excavating at Olduvai Gorge in Tanzania before they found what Louis wanted – a fossil species that combined ape and human characters, especially in the skull. In the process they did discover what were at the time the oldest known stone tools and many bones of extinct animals that lived alongside our extinct relatives.

Despite Louis's claims these also turned out to be more 'red herrings' and the next two decades were ones of considerable difficulty and turmoil, not only for Louis Leakey, but also for the investigation of human ancestry in East Africa – and of course there was the disastrous matter of World War II. Louis's first marriage broke up and he married Mary Nicol, a young woman who was very interested in his work and had some training as an archaeologist and illustrator. He first brought her to Africa in 1935 but it was not until the 1940s that they had enough funds to return to the business of serious field investigation.

More stone tools and many animal fossils were found. New sites were investigated and many important remains discovered such as that of the Miocene age ape *Proconsul* in 1942 on Rusinga Island, Lake Victoria. But frustratingly none of these were what most interested Louis – the fossils of early humans, those which clearly lay on the human side of Keith's 'cerebral Rubicon' with a brain big enough to encapsulate the emergence and development of human intelligence.

Of all the sites Mary and Louis investigated it still seemed as if Olduvai, with its abundance of stone tools, held out the best promise of finding the remains of those who made the tools. In those days it was still thought that the manufacture of stone tools was probably linked to an enlarged brain and human-type level of intelligence. The Leakeys returned to Olduvai in 1951 but the emergence of anti-colonial African political and social unrest, especially the 'Mau Mau Rebellion' in Kenya disrupted the work and it was not until 1959 that they could resume the search – and at last they struck lucky.

It was on 16th July, 1959 that one of their African field assistants spotted a large cheek tooth sticking out of a layer of sediment in the gorge. The next day Mary went to investigate and came back in great excitement, at last they had something to shout about and that would please Louis – they'd found a skull, broken into numerous pieces but retrievable. All it needed was time and patient excavation and restoration to recover and reconstruct the pieces – and Mary was very good at that.

The impressive skull was nicknamed 'dear boy' by Mary and 'Nutcracker Man' by Louis for the

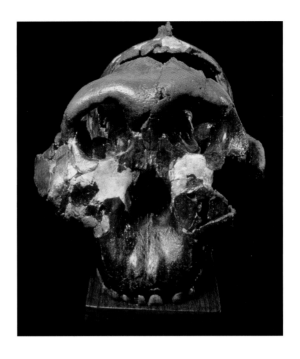

international press because of its massive muscular jaws. For the scientific community Louis coined the name *Zinjanthropus boisei* or 'Zinj' for short, with the species name referring to one of his main financial backers, the businessman Charles Boise. Louis thought that there was enough difference between Dart's australopithecines and his Olduvai find to

In 1959 the first good fossil skull of an extinct human relative was found at Olduvai. Although broken into many pieces it was reconstructed as an impressive robust australopithecine that Louis named as a new genus and species *Zinjanthropus boisei*. It is now generally known as *Paranthropus boisei*.

Several of the human related skulls found in South African caves such as Swartkrans have tooth and claw marks suggesting that they were preyed upon by big cats and raptors who carried the cadavers or severed heads into the caves to consume them at their leisure and safe from other predators or scavengers.

Palaeoanthropologists were initially reluctant to accept the results of radiometric dating of the Olduvai site.

A satellite image of the Main Ethiopian Rift Valley trending north-north-east into the Afar Depression towards Bodo and Hadar to the north. The flanking plateaux are uplifted older rocks whilst the Afar is a downfaulted depression with volcanoes and younger blocks of strata.

warrant placing it in the new genus *Zinjanthropus*, meaning 'man from East Africa' and he also at first thought that his new australopithecine was responsible for manufacturing the Oldowan stone tools. We now know, however, that this is unlikely.

Louis had examined Dart and Broom's Johannesburg collection of australopithecines and despite noticing the similarities between his find and theirs, persisted in claiming that the large cheek teeth of the Olduvai 'Nutcracker Man' warranted the establishment of a new genus. Although the find was still something of a disappointment to Louis because it was so small brained, he made the best of it in terms of publicity in Britain and then America in order to raise more funds for continued fieldwork. In the States he had the luck to 'hit it off' with the president of the National Geographical Society, Dr

Melville Bell Grosvenor, whose subsequent financial support made a huge difference to the Leakeys.

Louis also heard about the important progress that had been made in the technology of radiometric dating. His initial assumption was that Bed I at Olduvai might be around 600,000 years old but in 1961 an American team came up with the astonishing age of 1.75 million years old, virtually tripling the accepted estimate. Not surprisingly it was some time before the palaeoanthropologists were fully prepared to accept the new data.

Louis and Mary Leakey went on to make more important finds at Olduvai and these were to show that there was more than one kind of ancient relative within the human family 'bush'. However, it was their son, Richard, who made the next significant find of australopithecine remains in new territory

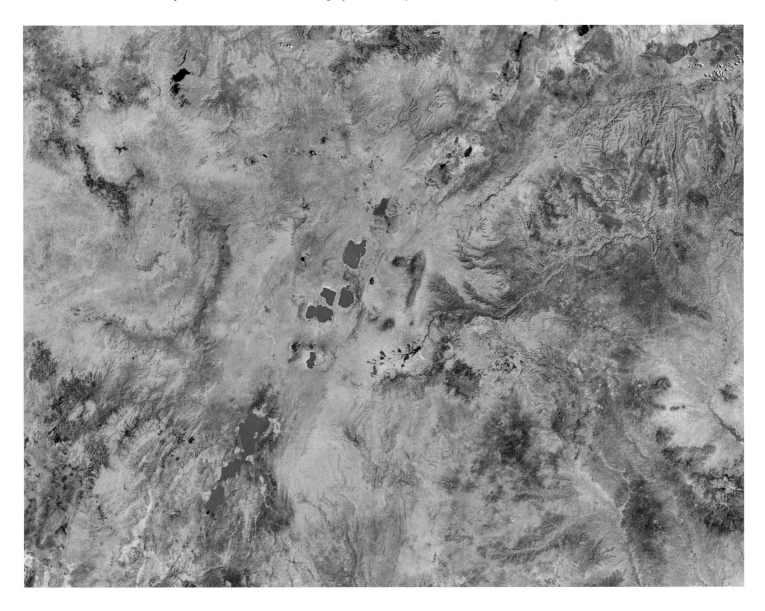

much further north on the eastern shore of Lake Turkana in Kenya. The strata here were known to contain numerous animal fossils of the right age – from late Pliocene up into Pleistocene times – and the horizontal strata are often well exposed around the lake and close to the surface in this region. So it is relatively easy to make rapid surveys of large areas of rock surface which might yield fossils of the animals and any of the much rarer human relatives who might have lived and died in the area.

In 1969 Richard Leakey was successful, finding a complete and very well preserved skull of a robust australopithecine, which could be assigned to the same species as 'Nutcracker Man'. However, by this time Louis' genus *Zinjanthropus* had been 'binned' as it was not significantly different from Broom's genus *Paranthropus*. Further finds to the north in the Hadar region of Ethiopia were clearly those of a different kind of human relative which was at first thought to be that of a new species of *Homo*.

The dramatic success of the Leakey family in finding fossil remains of our ancient relatives in East Africa and the attendant international publicity helped stimulate other research teams to 'try their luck' in Africa. French academic teams had been working in Africa for some time, especially in Ethiopia and Algeria but many of their members' time was constrained by teaching commitments back in France. It was the appearance of well-funded American academic research teams that was to make a significant difference and the whole business became considerably more competitive and complicated. At times it also became highly dangerous as civil war erupted in some countries such as Ethiopia.

In addition, and quite understandably, newly independent or emerging African nations became increasingly reluctant to give foreigners carte blanche for field research and then to remove important fossils to museums abroad. The Leakeys had always regarded themselves as Kenyans, given training to their African assistants and placed all their finds in what was originally the Coryndon Museum in Nairobi, later the National Museum. But, as we know the history of archaeology and similar fields is riddled with examples of national treasures being removed from the country of origin.

The 'Lucy' australopithecine skeleton

The arrival of Americans Donald Johanson and Tom Gray opened a new chapter in the story of human origins. In the early 1970s they discovered a spectacular skeleton in the Awash River area of the Hadar region of Ethiopia. Much of the geological groundwork in the region had been done by their French collaborators, Yves Coppens and Maurice Taieb, and another American geologist Jon Kalb. In 1973, just as Ethiopia was on the brink of a disastrous revolution, Johanson found some intriguing leg bones and then some jawbones, which he concluded had all belonged to a single human relative who walked upright, and who inhabited the earth approximately three million years ago. The following year came Johanson's 'grand coup' – the discovery of little 'Lucy'.

The Leakeys' success in Africa prompted others to follow – often into danger.

Lucy – The harsh and arid landscapes around the Awash River valley expose Pliocene age strata that have been scoured for human related fossils with great success. It was here that *Australopithecus afarensis*(Lucy) and the even older *Ardipithecus ramidus* have been found.

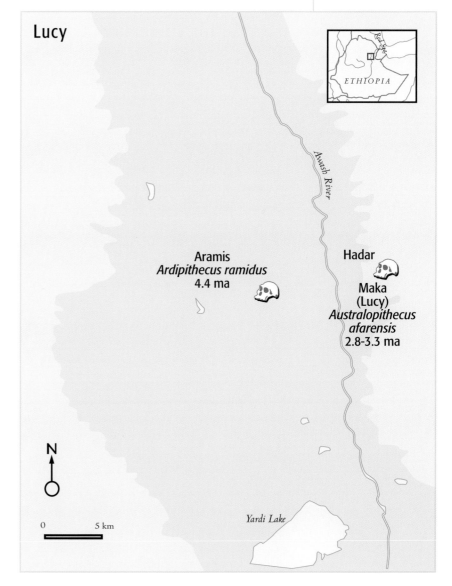

Lucy

ETHIOPIA

Awash River

Aramis
Ardipithecus ramidus
4.4 ma

Hadar

Maka
(Lucy)
Australopithecus afarensis
2.8-3.3 ma

N

0   5 km

Yardi Lake

Comparison of 'Lucy's skeleton with that of a chimp and modern human shows that although she could walk upright, she still retained a lot of ape features in her limbs and was perhaps as much at home in the trees as on the ground.

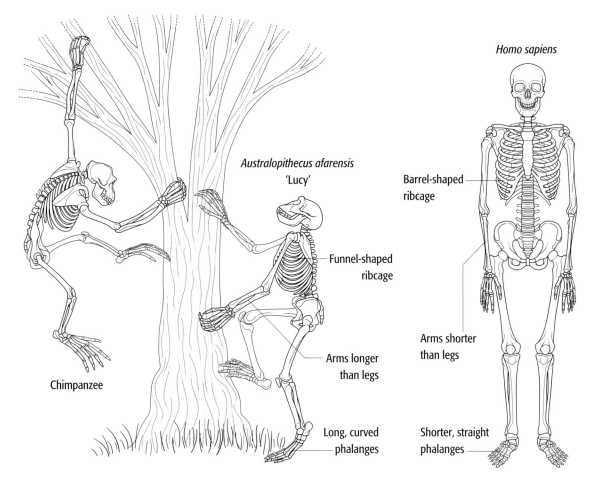

Homo sapiens

Barrel-shaped ribcage

Australopithecus afarensis 'Lucy'

Funnel-shaped ribcage

Arms longer than legs

Arms shorter than legs

Chimpanzee

Long, curved phalanges

Shorter, straight phalanges

As so often happens, the Laetoli discovery was made when the team wasn't even searching.

For the first time, a significant proportion of the skeleton (some 25 per cent, or 47 out of the 206 bones which make up the total skeleton) of an ancient relative was found with an estimated age of around 3.5 million years (subsequently revised to nearer three million). The major missing element was the skull but the lower jaw was present and it was quite different from that of the robust paranthropine australopiths, being much more lightly built.

The derivation of the nickname 'Lucy' is a well-known part of the folklore of palaeoanthropology. As the research team celebrated their success they were playing songs of the British pop group, the Beatles, and one particular song *Lucy in the sky with diamonds* took their attention so forcibly that they decided to name the find 'Lucy'. The find was an international hit, making headline news around the world – and Don Johanson an instant star on a par with the Leakeys.

Detailed analysis of the remains took some time, as did the formal assignment to a species. The V-shape of the jaw indicated a closer affinity with the australopithecines rather than with *Homo*, and study of the surviving pelvis bones supported the view that the skeleton was that of a female, so the nickname was appropriate. Most importantly, the leg bones confirmed what Raymond Dart had claimed nearly 50 years previously, namely the bipedal nature of these small-brained australopithecines. More detailed analysis of some of the foot and finger bones suggests that 'Lucy' and her kin still retained some tree-climbing abilities.

Otherwise a number of distinctly ape-like features are preserved in the skeleton. For instance, the rib cage is narrow in the upper part of the chest and flares out downwards, as it does in the living higher apes. This is to accommodate the large stomach and gut of these plant-eating animals. Plant food is difficult to digest and needs to reside for some time in the stomach so that bacteria and digestive enzymes can extract what little food value there is in a lot of plant material. By contrast the human rib cage is more tubular because of the smaller gut of our ancient human relatives who were typically omnivores or carnivores.

Footprints from the past

Laetoli lies some 50km (30 miles) south of Olduvai and in 1976 Mary Leakey and her team were exploring the possibility that this area might provide a new source of human-related fossils but were not having much luck. Out in the semi-arid bush some of the team felt the need of some light relief from the task of intensely searching the weathered rocks looking for any tiny speck of bone that might turn out to be important. While fooling around throwing African 'snowballs' (bits of dried animal dung) at one another, two of the team jumped down into a dry gully for cover – and stumbled across some footprints.

What immediately struck them was that the prints were not at the surface but were impressed *into* the surface of a rock layer beneath the soil. The rains that periodically flushed the gully had washed the soil cover away. Looking more closely they thought that the prints looked like elephant tracks and when they cleared more of the soil away they could see

The 1974 discovery in the Hadar region of Ethiopia of a partial australopithecine skeleton by Don Johanson and his team was a breakthrough. Although only 25% of the skeleton is preserved, it is very rare to find as many bones belonging to one individual although unfortunately the skull was missing. It was not until 1978 that the fossil was named as *Australopithecus afarensis*.

Laetoli – The Laetoli footprints lie to the south of Olduvai but still within an highly active volcanic region of the Rift Valley.

While Don Johanson and his colleagues were still considering the tricky problem of what to officially call 'Lucy' – her taxonomic position within the existing classificatory scheme of human relatives – another spectacular australopithecine-related find was made by Mary Leakey and her team down south in Laetoli, Tanzania.

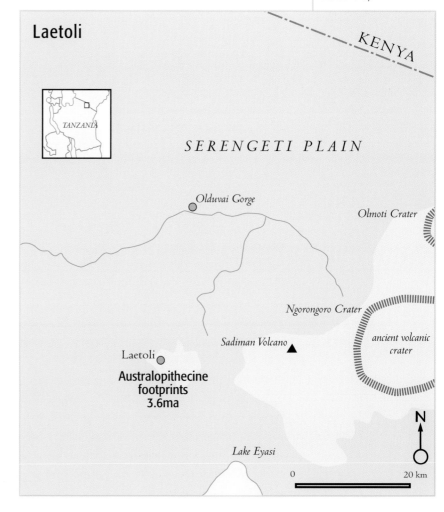

In 1978, Paul Abell, one of Mary Leakey's field team working at Laetoli in Tanzania discovered some human-like footprints impressed into a hardened rock surface and covered with soil. After some false starts that were probably animal tracks, the team had been looking for better preserved tracks and Abell found them. They rock was a hardened volcanic ash that was originally erupted by the nearby Sadiman volcano, some 3.6 million years ago.

that the tracks continued beneath the cover of soil and included the prints of other game.

Palaeontologists call such ancient trackways 'trace fossils' as they can record a variety of behaviour by the life of the past, which is not otherwise preserved in the fossil record. Mary Leakey was enthusiastic about the find and redirected the team's efforts towards uncovering more of the tracks. She knew that the rock layer preserving the prints was late Pliocene in age and that any information about the animals of the time would be valuable as they included many extinct forms. From a close examination of the rocks, Richard Hay, the team's geologist was able to reconstruct the sequence of events.

The animals had been crossing a newly fallen layer of volcanic ash from an eruption of the nearby volcano of Sadiman, which had spewed out clouds of ash, which had then 'rained' out of the sky blanketing the surrounding landscape. Within days of the fall, chemicals in the ash caused it to harden,

faithfully preserving the details of the various footprints impressed upon the surface. Then, fortuitously, further falls of ash and deposits of mud during flashfloods buried the surface in new layers of sediment and protected the prints from weathering and erosion. Radiometric dating of the ash later showed it to be around 3.6 million years old.

As the team excavated the prints they realised that there was the possibility of finding not just animal prints but those of ancient human relatives who might have been living in the area at the time. Altogether they found over 18 000 individual prints of a great variety (some 20 species) of animals, from rhinoceros to ostrich — and even insects. Then right at the end of the season they found some prints that did look quite human, but Mary was sceptical. However, by the end of the following season she was convinced that they were genuine human-related prints — and by far the oldest such footprints known. Most importantly, they indicated that whoever made them was an upright walking biped.

Mary eventually announced the find at a Press conference in Washington in February 1978, but the British anatomist Michael Day was still not convinced and told Mary so. Consequently at the conference Mary said that she was 75 per cent certain that they were human-related, but Day had sown seeds of doubt in her mind and she was determined to pursue the search.

Then, on 27 July, a young geochemist Paul Abell working with the field team found some new prints and showed them to Richard Hay who was impressed. When told of the new find back in camp Mary sent out Ndibo Mbuika, one of her most skilful assistants, to assess the find and he soon returned saying that there were indeed more prints of the same kind which could be uncovered.

This time everyone was convinced as there were two parallel sets of which the smaller trackway looked remarkably human. The large set was curious and not so clearly impressed, but even Mary, who was normally pretty phlegmatic, having experienced too many false hopes, had a big grin and kept on saying 'Oh, we found it, we found it'.

Indeed they had. By the end of the season they had uncovered 23m (75ft) of the double trackway, which preserved over 20 consecutive steps made by a

It seemed as though one of the beings stopped for a moment and turned... a look back, in time.

When the Laetoli footprints were finally uncovered they revealed a mass of animal tracks and a wonderful parallel set of very human-like footprints. Two fully bipedal individuals, one bigger than the other, walked slowly and perhaps side by side across the soft ash. The prints conclusively proved that upright walking human relatives had evolved by this time as the fossil bones indicated.

The site was covered with sand and stones for protection but plants grew through the sand. A new and more thorough conservation plan has now been put in place.

couple of bipeds as they walked across the soft ash. There was absolutely no doubt that the walkers who made the tracks were fully bipedal in the modern sense. Although chimpanzees can walk bipedally when they want to, they waddle in a similar way to a human toddler. They have to swing the body from step to step as the leg joints with the pelvis do not bring the legs right under the centre of gravity. As no members of the genus *Homo* had yet evolved 3.6 million years ago, the prints had to have been made by an australopithecine. Again it was evident that Dart had been right, not only had 'Lucy's' leg bones confirmed it, but the footprints clinched it. A large human-type brain is not needed to walk upright and so adaptation to bipedalism was promoted by some other force, the question was what?

The Laetoli find was published in *Nature* in 1979 and pictures of the tracks quickly found their way into national newpapers and journals around the world. The National Geographical Society of America was still sponsoring Mary and at their press conference she could not resist the temptation to speculate about the original circumstances. It seemed as if two australopithecines, one much larger than the other, had walked side by side – and it even looked as if one had briefly hesitated and turned before carrying on. The publicity attached to the find provided funds for further investigation and Mary assembled an international team of investigators for the following field season.

The parallel tracks were not only in step, but very close together and the bigger set gave the impression of a curious second strike as if there was a third individual who carefully 'dogged' the large individual step by step. This interpretation led to all sorts of romanticizing with the tracks being made by a 'family' – male and female arm in arm with an infant behind walking in father's steps. It is a nice story but there is no evidence for it as we do not know what sex or age the individuals were, nor whether they were all together when the prints were made.

In order to preserve them, the trackways were covered up at the end of the 1979 field season. Re-excavation in 1995 showed that some damage had been caused by growth of acacia shrubs and so the tracks were more carefully protected before being reburied in August 1996, just months before Mary Leakey died. Prior to being covered, a cast of the

The picture was to change as more ancient relatives were discovered.

prints was made and they can be seen in a permanent exhibit in the small Olduvai museum on the road to the Ngorongoro Crater, overlooking the gorge where Louis Leakey began his African work over 60 years previously in 1931.

Giving 'Lucy' a name

The Laetoli find was a spectacular discovery, and the story just related has taken us from their discovery in 1976 up to 1996. In 1978, meanwhile, Don Johanson named 'Lucy' as a new species of australopithecine – *Australopithecus afarensis* – but in doing so he greatly upset Mary Leakey because he chose one of her specimens from Laetoli – a jawbone with nine teeth – as the type specimen.

Mary had been reluctant to give the jaw a new name until she had more and better material for the species. Now Johanson had done just that in the 'Lucy' skeletal material – which had in fact come from Hadar in Ethiopia over 1600 km away. Not only that but the Laetoli specimen is somewhat older at somewhere between 3.5 and four million years old. Johanson had offered to make Mary co-author of the new species but she had refused as she disagreed with the naming of a new species based on material from such different sites. She was also annoyed that the species name should be *afarensis*, referring to a region in Ethiopia, when the critical jawbone came from Laetoli in Tanzania, but there was nothing she could do about it.

Australopithecus afarensis predated Dart's *A. africanus* and for nearly 20 years was the oldest known australopithecine and seemed to hold a pivotal role in the evolution of all subsequent human relatives. It seemed that somehow or other both the robust and graceful australopithecines and the earliest members of the genus *Homo*, all of which appeared over the next million or so years had to be derived from 'Lucy' and her kin. However, as we have seen, the picture was to change when more ancient relatives were discovered and, as we shall see, when the origin of *Homo* came to be reassessed and questioned.

Within the australopithecine context, 'Lucy' was displaced from her 'ancestral' position when, in 1995, Meave Leakey (wife of Richard Leakey) and a team who had been working at Kanapoi in Kenya, announced the naming of a new and older australopithecine – *A. anamensis* (4.2-3.9 million years old). The new species was based on an adult jawbone and teeth but more recently a partial skull of a young individual and some limb bones have been discovered. The latter indicate that like 'Lucy', this earliest known australopith was also capable of walking upright.

Today, the diversity of australopithecines might seem surprising but in their 'bushy' evolution they are following a repeated pattern seen not only throughout the primate mammals, but in the mammals as a whole and even on a much larger scale within the evolution of all past life. These little metre or so high extinct human relatives had acquired a very successful adaptation – bipedal walking which allowed them to spread over a considerable part of Africa. Populations became separated by changing geographical and environmental barriers and evolved in isolation into a number of distinct species. One group, the robust species such as *Paranthropus robustus*, became adapted for feeding on tough fibrous plant materials (such as roots, tubers and perhaps some fruit and foliage) while the more gracile species such as *A. afarensis* probably became more omnivorous but still relied mostly on a plant based diet (perhaps fruit).

Interpretation of the plants and other animals which lived alongside *A. afarensis* in the Hadar region shows that the environment was radically different from the harsh arid landscape that it is today. Around three million years ago there were lush lakeshore marshes and riverbanks with a wonderful diversity of aquatic and terrestrial plants and animals.

It is also possible that *A. afarensis* began to exploit the availability of meat protein just as chimpanzees occasionally do today. However, it is unlikely that they actively hunted anything but small game and probably relied on scavenging from kills made by other more powerful predators such as the big cats. Like hyenas, the australopithecines may have been able to drive a big cat away from its kill by co-operating as a group and harassing it. The advantage of the meat protein is that it would have helped fuel the increased energy required in walking any distance and it would have supplemented a plant-based diet in times of drought or in more open savanna grasslands where edible plant food is scarce. As we

shall see, access to and use of meat proteins may have brought other benefits as well – fuelling brain development.

Bipedalism brings a number of benefits, primarily the freeing up of the hands for other tasks such as carrying food, infants and tools. The down side is that in order to walk upright efficiently the feet, ankles, legs, knee and hip joints have to be modified in such a way that results in them no longer being so efficient for climbing trees. This is a typical evolutionary compromise and it meant that for the australopithecines they would have lost some of the safety provided by adept tree climbing in return for enhanced ability in walking upright across open spaces. While the upright stance provided better sight lines than those provided by a quadrupedal knuckle-walking stance most of their predators would easily have been able to out run them in the open. So that some other strategy was probably employed to maximize safety such as group vigilance and maybe the use of sticks and stones as weapons.

This collection of ancient human remains comes from the region around Lake Turkana and and was found by Richard Leakey's team. It includes (from right to left) skulls of *Australopithecus boisei, Homo ergaster* and *Homo rudolfensis*.

CHAPTER SEVEN

Mind the gap – searching for the roots of the family 'bush'

The biggest breakthrough in the search for our ancestry has been the discovery of a few African fossils that lie right at the base of the human family tree. For decades there was a huge lacuna between the oldest human related fossils — australopithecines not much more than three million years old and their nearest primate ancestors — fossils such as *Proconsul*, some 23 million years old. The molecular clock suggests that we split from the apes between seven and five million years ago. So where were the fossils of the roots of the family 'bush' that could fill the gap?

Five skulls belonging to some ancestors and relatives of modern humans. They are (from left, clockwise): *Australopithecus africanus*, 3-1.8 MA; *Homo habilis* (or *Homo rudolfensis*), 2.1-1.6 MA; *Homo ergaster*, 1.8-0.3 MA; *Homo sapiens*, 92,000 BP; Cro-Magnon human (a later form of *Homo sapiens*), 22,000 BP.

As we have seen, the pioneers of human evolution, Thomas Henry Huxley and Charles Darwin, clearly showed the close biological relationship between humans and higher apes. Darwin in particular went further and predicted that the evolutionary ancestry of humankind was to be found in Africa where the gorilla and chimpanzee live. Although Darwin's great disciple the German biologist Ernst Haeckel subsequently argued that the gibbons are closer to humans and that therefore the 'Eden' of humanity should be closer to Asia, we now know that he was wrong on both counts.

Biologically and genetically, our closest living relation turns out to be the chimpanzee, for although Haeckel's protégé Eugene Dubois' discovery of 'Java Man' (now known as *Homo erectus*) in the 1890s seemed to vindicate his hero's theory of an Asian origin for humankind, that also turned out to be a wrong turning in our understanding of the human story. Today, Africa is generally accepted as the 'Eden' or cradle of humanity. However, there is still an Asian part of the story relating to the *Homo erectus* people that is not clear. However this does not detract from the overwhelming evidence for *Homo sapiens* originating in Africa.

The most forceful argument for that African origin has come from the genetic analysis and comparison between the living higher apes – especially the chimpanzees – and us. With more than 98% of our DNA in common, not only are we very closely related but that closeness also implies a remarkable recency in the evolutionary split between the two groups. According to the molecular clock that division occurred somewhere between five and seven million years ago.

The problem has been that until recently there was virtually no fossil record of our most ancient human-related ancestors from either side of the great divide. However, in the last decade or so some remarkable discoveries from new sites in Africa have at least helped fill in the great gap in the fossil record between four and seven million years ago and radically changed the emerging picture of our ancestry. The new sites include some which are over 2,500km (1,600 miles) to the west of the Great East African Rift Valley, which hosts most of the fossil remains of our fossil ancestors and their relatives.

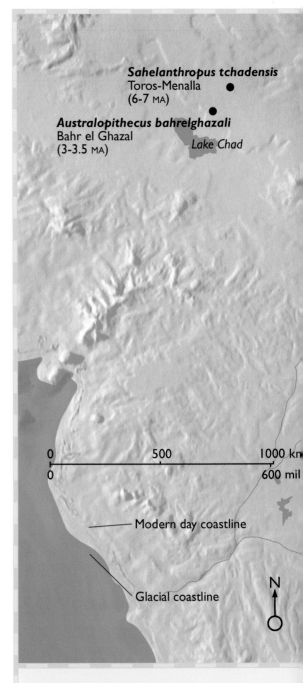

Sahelanthropus tchadensis
Toros-Menalla ●
(6-7 MA)

Australopithecus bahrelghazali
Bahr el Ghazal
(3-3.5 MA)   *Lake Chad*

0   500   1000 km
0   600 mil

———— Modern day coastline

———— Glacial coastline

N

Deep ancestral roots

Africa is gradually revealing the deeper roots of the human family tree. A few decades ago the most ancient fossil ancestor was around 3 million years old and beyond that there was a huge gap in the fossil record back to the Miocene apes. The gap was both temporal and anatomical and the hopes of finding any fossils to fill it seemed to become increasingly forlorn.

This was partly because it was difficult to find well exposed strata of the right age ie between

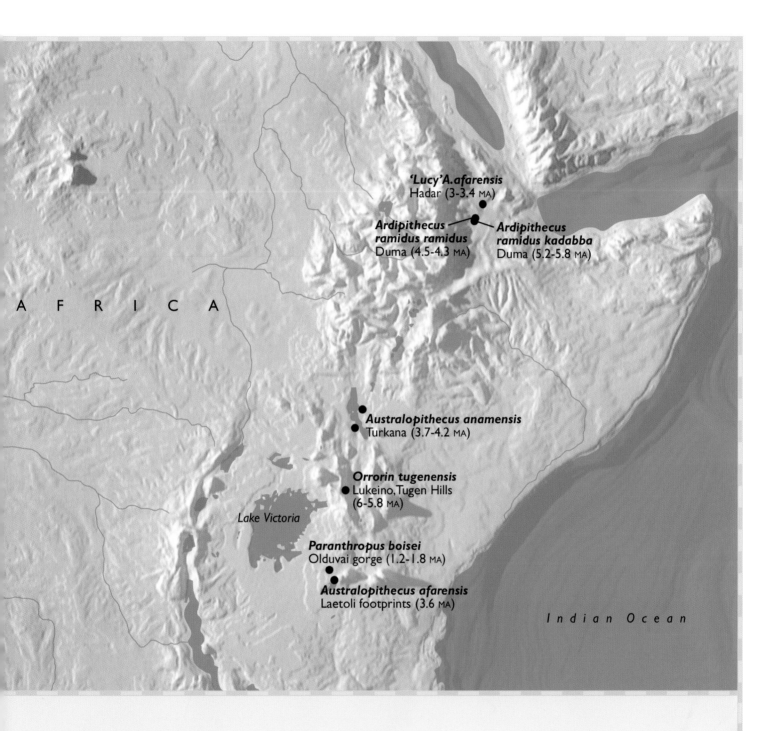

'Lucy' A. afarensis
Hadar (3-3.4 MA)

**Ardipithecus
ramidus ramidus**
Duma (4.5-4.3 MA)

**Ardipithecus
ramidus kadabba**
Duma (5.2-5.8 MA)

A F R I C A

Australopithecus anamensis
Turkana (3.7-4.2 MA)

Orrorin tugenensis
Lukeino, Tugen Hills
(6-5.8 MA)

Lake Victoria

Paranthropus boisei
Olduvai gorge (1.2-1.8 MA)

Australopithecus afarensis
Laetoli footprints (3.6 MA)

Indian Ocean

4 and 10 million years old. Eventually, as the geological mapping of the huge region of Central Africa was able to point to outcrops of strata of the right age, especially in the Afar region of Ethiopia and Kenya's Tugen Hills, both within the East African Rift region. Afar produced the new genus *Ardipithecus* and the Tugen Hills another one *Orrorin*.

The assumption was that early human relatives had probably been restricted to this fertile region of lakes, rivers and woodland with abundant and diverse wildlife.

But it was also clear that such woodlands extended well beyond the Rift, as did the accompanying game. Persistent searching the inhospitable desert region of Chad, well to the west of the Rift eventually proved extremely worthwhile with the discovery of the oldest known human relative – *Sahelanthropus tchadensis*. The latter has been dated by long distance fossil correlation to between 6 and 7 million years old. According to the molecular clock, the estimated timing of the divergence between the ancestral chimp and human lineages coincides with this part of the Pliocene. *Sahelanthropus* may be very close to that common ancestor, although some experts argue that it is on the ape side of the divide.

Toumaï – *Sahelanthropus tchadensis*

Most dramatic of all has been the discovery of a remarkable well-preserved skull, *Sahelanthropus tchadensis*. This extraordinary fossil has been nicknamed Toumaï, which means 'hope of life' in the local Goran language of the north central African territory of Chad where it was found. Today, this region is part of the blisteringly hot Saharan desert lying between the Tropic of Cancer and the Equator.

Michel Brunet's wonderful skull of *Sahelanthropus* was quite badly distorted when it was found but careful reconstruction has shown that it has a mosaic of ape-like and some human-like features.

Right: Michel Brunet admires his new find – *Sahelanthropus.*

A combined French-Chadian team led by Michel Brunet of Poitiers University has spent over 10 years here searching for the fossil remains of our ancestors. In upper Miocene sedimentary deposits, which poke through the desert sands, they have found thousands of fossils of some 44 different kinds of extinct animals that occupied the region seven million years ago.

Altogether, the fossils – which range from extinct kinds of fish and crocodiles to rodents and elephants, along with their surrounding deposits – show that in late Miocene times the region enjoyed a radically different climate and landscape from that of today. It would have comprised swamps and forest combined with more open savannah grassland, perhaps somewhat like the Okavanga Delta of Botswana today. Nevertheless, sandy desert was not too far away and marked the beginning of the climate change with increasing aridity that was going to radically affect the subsequent history of life in Africa, and perhaps the development of our ancestors.

Within the sedimentary sequence of upper Miocene age deposits, one particular layer of sandstone contains most of the remains of the land-living animals. The palaeontologists call it the anthracotheriid unit after a large, extinct anthracotheriid called *Libycosaurus petrochii*, the fossil

Two versions of a *Sahelanthropus* reconstruction.

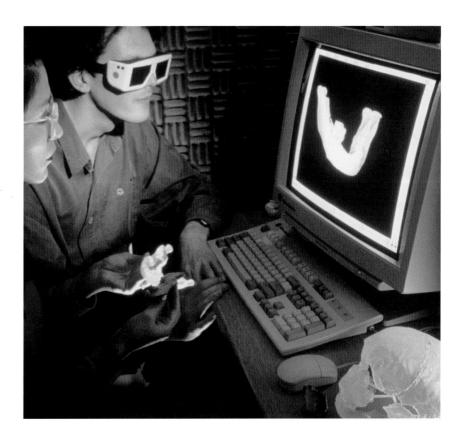

remains of which are quite common. It was a semi-aquatic mammal a bit like a hippo but with very different head shape and it lived alongside primitive loxodont elephants and suids (pigs) like *Nyanzachoerus syrticus* that foraged on the lush vegetation. The predatory and scavenging carnivores included hyena-like and large cat-like animals. In addition there must have been raptors and many other kinds of birds, but so far their fossil remains have not been found. In among this typically diverse tropical game 'park' lived a small metre-sized and distinctly ape-like human relative – *Sahelanthropus tchadensis*.

Discovered in 2001, the remains primarily consisted of a nearly complete skull plus a few jaw fragments and isolated teeth. Although the skull was in one piece, it has been crushed and distorted by the post-mortem process of burial. Nevertheless, Brunet and his team detailed its features and claimed that although many were indeed distinctly ape-like there were other features which suggest that not only does it lie on our side of the human-ape split but that it might have walked upright. The ape-like features include its low-domed and elongate skull shape, small brain size (just 360-370cc in volume) and very prominent bony brow-ridge, behind which lies a very marked 'waist' or narrowing which is best seen from above.

Among more advanced features are the small size of the teeth at the front of the mouth, especially the canines, and some of the angular relationships between parts of the brain case and the short face. All in all, the skull shows a unique, patchy mixture of primitive and slightly more advanced features.

Counter-claims

No sooner had the description been published than critics presented a counter-claim that *Sahelanthropus tchadensis* represents no more than an ancestral ape, perhaps a female gorilla ancestor, and suggested that its name should be changed to *Sahelpithecus tchadensis*. Similar criticisms were made in 1925 when Raymond Dart first described his then new genus *Australopithecus*. But in the intervening years more finds of teeth and jaw fragments have been made.

The new fossils represent the fragmentary remains of at least six and perhaps as many as nine individuals from three adjacent sites among which there is a lower canine tooth with an advanced non-honing surface. By contrast chimpanzee upper and lower jaw canines typically abrade against one another (occlude) to produce a self-sharpening wear surface. In addition the enamel thickness (1.2-1.9mm) of the cheek teeth is intermediate between that of the chimpanzees and the australopithecines.

Recently, a sophisticated reconstruction of the skull has been made using the latest high-resolution computerized tomography (CT) based on a series of CT scan 'slices' taken every 0.4mm through the skull. As reconstructed, the undistorted skull is claimed by the authors to show a number of features that verify its position as the earliest known human-related ancestor. The face and floor of the braincase were separately analyzed and evaluated as these are critical features in helping to indicate their evolutionary status on the chimpanzee-to-human progression, and whether the animal was capable of walking upright or whether it habitually used knuckle-walking as in the surviving higher apes. Sophisticated modeling techniques were employed to test whether the skull fragments would best fit a *Pan*-like (chimpanzee), *gorilla*-like or australopithecine-like facial configuration. The authors claim that the best fit is with the latter and that this can be supported with statistical tests.

Computer aided tomography allows the three-dimensional manipulation of skulls and their remodelling to remove postmortem distortion.

Climate change has often had a radical impact on our history, and influenced what we are today.

Recent research appears to verify *S. tchadensis*'s position as our earliest known human-related ancestor.

Remodelling of the *Sahelanthropus* skull to remove its distortion shows that the angle between the facial optical planes (OP) and the foramen magnum for the central nerve cord (FM) is more human than ape-like and may indicate that it was capable of walking upright.

A computer image of a Neanderthal child's skull shows it large brain and lack of a bony browridge at this stage in development.

In profile, the angle and forward projection of the reconstructed face is significantly different from that of a chimpanzee, and closer to that of an australopithecine. However, the relative size and shape of the brain case is less inflated than that of an australopithecine. In addition, considering that *Sahelanthropus* is well over two million years older than the oldest known australopithecine, it has a remarkably advanced face. It might reasonably be expected that its face should have been like that of a more primitive ape. Given its age and closeness to the common ancestor between the chimpanzee and human lineages, there could well have been a greater forward projection of the muzzle region and presence of a bony crest for muscle attachment on the top of the skull – but there is not.

The angular relationship of the foramen magnum and the near vertical plane within which the eye sockets are set (the orbital plane) are similar to those of the australopithecines. The foramen magnum is the hole through which the central nerve cord passes from the brain stem down into the body. In upright-

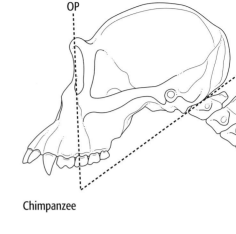

Chimpanzee

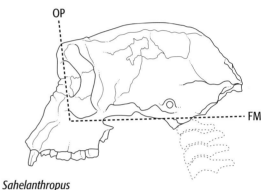

Sahelanthropus

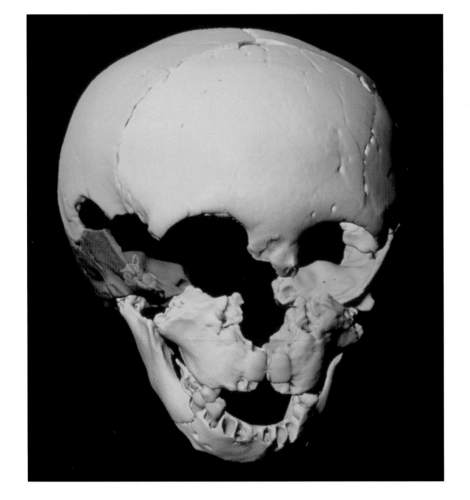

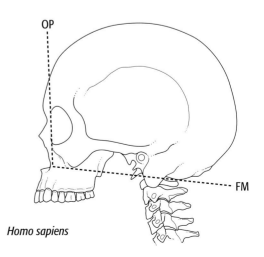

Australopithecus africanus

Homo sapiens

walking human relatives the hole is placed on the floor of the braincase so that it exits close to a right angle with the orbital plane. By contrast, in the knuckle-walking chimpanzees and other quadrupedal mammals whose backbone is typically parallel to the ground surface and yet whose vision is directed parallel to that ground surface, the hole is placed at the back of the cranium. This is the same kind of evidence that Raymond Dart used to predict that the australopithecines were bipedal.

Altogether, *Sahelanthropus tchadensis* preserves this curious mixture of primitive chimpanzee-like and more advanced australopithecine features and may well have been an upright walker. The final proof however, will lie with the recovery of some of the critical parts of the leg bones, especially the articulating surfaces of the joints, as was discovered in 'Lucy' (*Australopithecus afarensis*) by Don Johanson and his team back in the 1980s when they confirmed Dart's bipedal claim.

So far, it looks as if *Sahelanthropus tchadensis* has become accepted as a genuine member of the human family 'bush', although if its limb bones turn out to indicate that it did not walk upright that welcome might weaken. At the moment one of Toumaï's most important features is 'his' age. But it is worth while just digressing a moment to consider how the age of around seven million years has been arrived at.

How do we know Toumaï's age?

The fossil remains of Toumaï did not come with a convenient name-tag or a date of birth. In fact, dating the fossil remains proved something of a headache for Brunet and his team. Many, if not most of the bones we have been discussing have been indirectly dated by radiometric methods applied to certain kinds of rocks adjacent to the fossils. Most of these radiometrically datable rocks are essentially igneous in origin and often are volcanic, especially within the context of the Great East African Rift Valley. However, the geological context of the *Sahelanthropus tchadensis* find from Chad is far removed from any volcanoes that were active in late Miocene times, so there was no possibility of dating the finds radiometrically, and the scientists had to fall back on the more traditional method of relative stratigraphical dating through correlation of fossils and the sedimentary rocks in which they were found.

It is the association of Toumaï's bones with those of the extinct mammal species, which provides the essential key to its relative age, however. The cheek teeth of the pig *Nyanzachoerus syrticus* from Chad are very similar to suid teeth found in East Africa's Lothagam Nawata sedimentary strata, which are between 5.2 and 7.4 million years old. Suids might be an unglamorous group of animals, but they were and still are a very successful and common group of mammals that have evolved quite rapidly through many different forms (genera and species) over the last few tens of millions of years. Consequently, their fossil remains are often quite abundant, and fortunately individual species can be distinguished just from the shape of the teeth. As a bonus, these are often the best-preserved fossils because, having no food value, the animals were of little interest to scavengers.

Over many years, palaeontologists have been able to work out how the succession of interrelated suid and other mammal taxa are distributed through ancient layers of sediment, and thus how their existence related to time. Associations of particular species are now known to characterize successive biozones, which in some places, such as the Tugen Hills and Lothagam in Kenya, can be related to the radiometric timescale through the occurrence in rocks of datable minerals such as volcanic ashes or lavas, which are interbedded with the sediments.

By finding identifiable and biochronologically significant fossils alongside the human-related remains, the scientists were able to say that their fossils belong within a particular known biozone. By correlating between this zone in Chad and the same zone within a radiometrically dated succession elsewhere, the scientists are able to reasonably conclude that the human related fossils belong within a particular age range, which in this case is between six and seven million years ago.

How did they live?

Our ancient *Sahelanthropus* relatives were, it seems, very ape-like and relatively vulnerable creatures. Even if they could walk upright they would not have been capable of running and when on the ground – and especially when in open spaces – they would have been particularly vulnerable to fast-moving predators. Like chimpanzees they may have been able

As the relics are pieced together, so a more complete picture will emerge.

The discovery of fossils of human-related species within a particular biozone allows scientists to accurately relate fossils that are sometimes considerable distances apart.

to move quite quickly using a knuckle 'walking' 'gallop', but they do not seem to have had such well-developed canine teeth as chimpanzees for self-defence. They probably relied on group vigilance and well-developed senses of sight, hearing and smell to detect danger.

Most likely they were plant or fruit-eaters and spent most of their time in the treetop canopy, safe from most predators. They probably would have spent as little time as possible on the ground. Important questions about how they got around – whether they were knuckle-walkers or had arms, wrists and hands adapted for swinging from branch to branch remain unanswered until the appropriate bones are found. Their brain capacity was certainly no more than that of chimpanzees, but as we now know, chimpanzees can manage to lead successful social lives and show considerable intelligence (see page 28).

By not placing the Toumaï fossils in any previously described genus and giving it the new genus name (*Sahelanthropus*), meaning 'human from the Sahel', Brunet and his team have thrown down a taxonomic challenge. They are effectively saying that their new human relative is distinctly different from any previously described member of the human family 'bush' and consequently leave open the question of how it is related to all of the already known members.

Experts remain at odds over the interpretation of the characteristics of the *O. tugenensis* femur head.

The discovery in 2000 of fossil leg bones, nearly six million years old, in the Tugen Hills of northern Kenya has led to much argument between experts. Brigitte Senut and Martin Pickford who found the fossils claim that their new species, *Orrorin tugenensis*, could walk upright but other experts disagree.

One of the most important aspects of the Toumaï discovery is its unique assemblage of characters. It is different from that seen in the chimpanzees and all other fossil relatives. They remind us that single characters or sets of characters can evolve at different rates and some of them which are closely related to function may evolve at different times in different groups which need not necessarily have a direct evolutionary relationship.

We are only just seeing the beginning of the Toumaï story and the assessment of its role in our ancestry. Now, however, we need to assess the potential role of the next candidate in our chronology of putative ancestors – another fragmentary species from Africa: *Orrorin tugenensis*.

Orrorin tugenensis (6-5.8 million years ago)

Found in 2000, it was inevitable that the fossil remains were initially nicknamed 'Millennium Man' but now over five years later the name no longer seems appropriate. The remains consist of a few fragments of limb bone, a few teeth and a jaw fragment. The discovery was made in the Tugen hills of northern Kenya by another French-based team led by Brigitte Senut and Martin Pickford. Most importantly they claimed that the form and structure of the ball joint that is preserved on the end of one of the leg bones indicates that the owner was capable of walking upright (bipedal locomotion).

Named *Orrorin tugenensis*, meaning 'original man from the Tugen Hills', the find has been dated to between six and 5.8 million years ago, still within late Miocene times. As in *Sahelanthropus* the teeth have relatively thick enamel but the canine is more distinctly ape-like than that of *Sahelanthropus*. The exact nature of the critical femur head has been a matter of some dispute between experts, however.

Pickford and Senut claim that the external form does show indications of bipedalism but other experts say that the difference is not significant enough to make that distinction. Another possible criterion is the internal wall thickness of the neck between the ball joint and the leg bone. This has been determined non-destructively by using a medical CT (computerized tomographic) scan. Again the interpretation of the scans is disputed.

Ardipithecus (5.8–4 million years ago)

As we move up into slightly younger strata and across the boundary from the end of the Miocene (5.3 million years ago) and into the beginning of Pliocene times, another two finds from further north in Ethiopia have to be considered. Both have been placed in yet another genus, this time called *Ardipithecus* meaning 'ground ape'.

Two species have been distinguished with *A. kadabba* the older (5.8-5.2 million years old) and *A. ramidus* (4.4 million years old) the younger. Although more fossil bones and teeth have been found and again there is a claim of bipedalism, there is still dissension among the experts as to whether this is really so. *Ardipithecus ramidus* was found over a decade ago now and the technical details have still not yet been published, apparently because the bone material is so fragile and fragmentary that it has required a very lengthy conservation process. The relics may yet provide more conclusive proof of bipedalism, however.

Although these three most ancient possible human relatives have all been placed in separate genera, there are aspects of the teeth that show considerable similarity. Indeed it is possible that they are all different species of a single genus, but until more and better fossils are recovered it is better to separate them at the generic level. At least their discovery shows that it is possible to find fossils within what was thought to be an interval of time bereft of such human-related material. In addition, they show that such fossils can be found in strata well beyond the Great East African Rift Valley.

It may well be just a matter of time before the search for further fossiliferous sedimentary deposits of the right kind pays off. Part of the problem is that the original population numbers of these human relatives were probably very low and the chances of much skeletal material being preserved is also very low compared with the more robust and common bones of larger mammals such as bovids (cattle relatives).

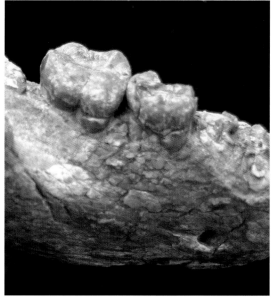

Ardipithecus ramidus is yet another ancient human relative from East Africa. Over 5 million years old, it is also thought to have been bipedal but some of the remains are so fragmentary, it has taken many years of painstaking work to reconstruct them.

This jaw fragment of *Ardipithecus ramidus* from Gona is one of the largest bits of this species found; all the rest is even more fragmentary.

People from different ethnic groups show biological adaption to their environment, through skin tone for example. The differences are only skin deep though, and we are all members of the same species.

PART III

Perhaps the most remarkable and most puzzling part of our story is the when, how and why of our global dispersal beyond Africa — a mere 100,000 years ago.

Our ancestors' diaspora beyond Africa is of particular interest to us today because it concerns the interaction of climate and environmental change upon our ancestors, and how our ancestors impacted upon their environment.

CHAPTER EIGHT

The dispersal of humans and the development of languages

 The global human population of some six billion is dispersed across the Earth from pole to pole and over the millennia has become adapted to a vast range of environments and climates, from hot deserts to alpine and polar tundra to tropical rainforests. Consequently, it is not surprising that we humans vary in a number of our most obvious characters such as body size, shape and skin pigmentation and less obvious ones such as blood type. Due to small genetic changes separate populations also show variations from the global norm such as in their tolerances to different environmental conditions and in their susceptibility to different diseases.

The full complement of male chromosomes is shown here, in a coloured light micrograph. Each human cell contains 46 chromosomes, 23 of maternal origin and 23 of paternal origin.

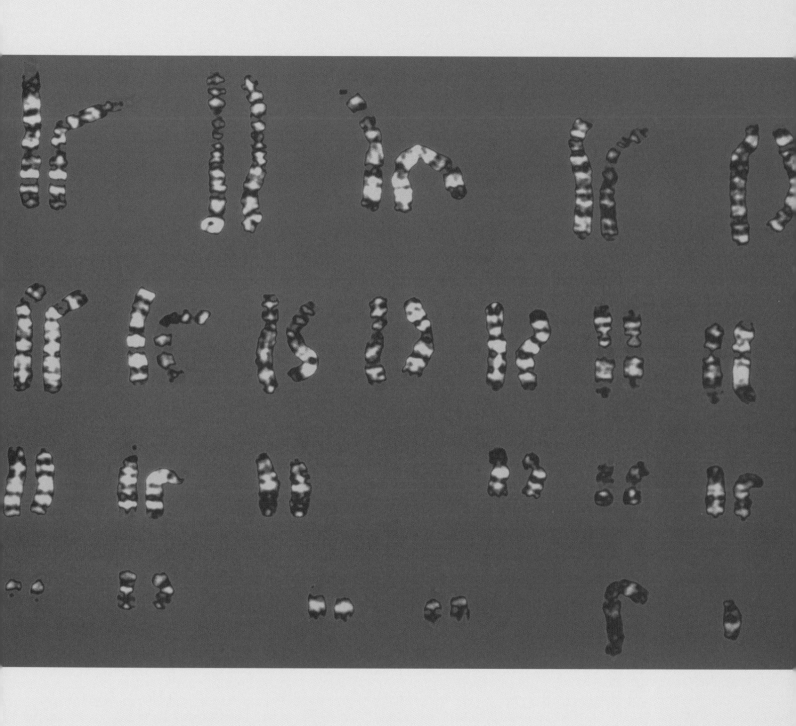

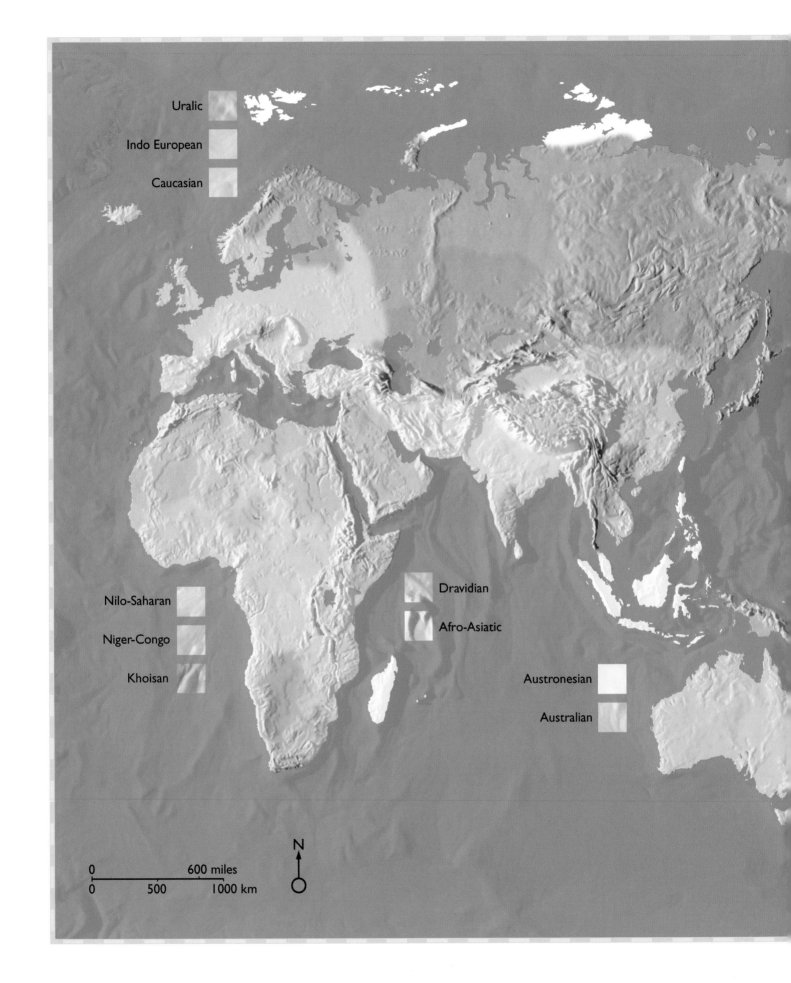

Uralic

Indo European

Caucasian

Nilo-Saharan

Niger-Congo

Khoisan

Dravidian

Afro-Asiatic

Austronesian

Australian

N

0   600 miles
0   500   1000 km

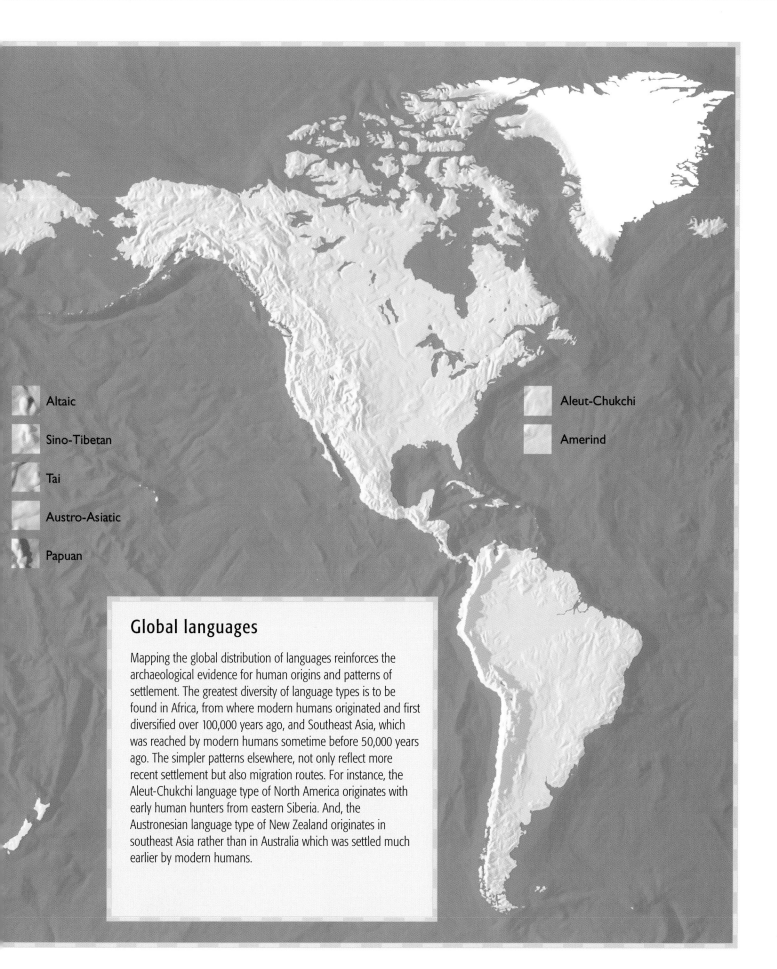

Altaic

Sino-Tibetan

Tai

Austro-Asiatic

Papuan

Aleut-Chukchi

Amerind

Global languages

Mapping the global distribution of languages reinforces the archaeological evidence for human origins and patterns of settlement. The greatest diversity of language types is to be found in Africa, from where modern humans originated and first diversified over 100,000 years ago, and Southeast Asia, which was reached by modern humans sometime before 50,000 years ago. The simpler patterns elsewhere, not only reflect more recent settlement but also migration routes. For instance, the Aleut-Chukchi language type of North America originates with early human hunters from eastern Siberia. And, the Austronesian language type of New Zealand originates in southeast Asia rather than in Australia which was settled much earlier by modern humans.

Virtually all species vary in some way or another – because of the nature and mechanisms of sexual reproduction. Only asexual reproduction minimizes the variation in a species. Furthermore, there is variation at other levels from within separate populations and even within families. We are a social species that emphasises and recognises both difference and similarity at all levels. Indeed we expect offspring to look like their parents and are suspicious if they do not.

And yet, despite all this variation we are still a single species that can interbreed and produce viable young who can in turn interbreed. Despite some thousands of years of physical separation of some populations we still share so much of our genetic material that there is no biological barrier to our interbreeding.

Indeed, it is the measurable commonality of our genetic makeup that is the main support for the palaeoanthropological evidence of our recency as a species – somewhere between 50,000 and 120,000 years ago in Africa. Mapping of the human genome and comparison with those already published, such as for the mouse and even more importantly with the chimpanzee, which is still in progress will provide a basis for extensive comparative studies that should shed light on how we differ from the

chimpanzees and when these changes might have occurred. The recovery of ancient biomolecules such as DNA from extinct relatives such as the Neanderthals is also providing evidence of past inter-relationships. In the future, mapping of other higher ape genomes especially that of the gorilla will provide an even better measure of the timing and rate of those changes that have made the difference between being an ape and being a human.

The study of genetics is one of the most important of modern scientific endeavours and the study of human genetics is most pertinent to our efforts to understand ourselves as a species with a long evolutionary past – and hopefully a future. Working in the mid 19th century, Darwin did not have the benefit of knowledge of genetics, but he realised that an understanding of the mechanisms and modes of inheritance would provide a key to evolution. We humans have interfered with the breeding of plants and animals for thousands of years, domesticating many species for a variety of purposes, but mainly for food, clothing and pleasure both sporting and aesthetic.

Darwin was very familiar with the breeding of domestic animals and what could be achieved by selecting potential mates that 'sport' particularly desirable traits ranging from better wool or meat to fancy plumage. He bred pigeons because they reproduce easily in captivity and had already been artificially developed into a number of distinctive breeds such as the tumbler (which has a characteristic flight path) and pouter (which has an inflated 'chest'). He found that despite all the exotic traits they did not always breed true and could easily be reverted to a form that approximated to the common pigeon from which they had originated a few hundred years previously.

We humans have also interfered with our own heritage and not just at the level of personal choice of a mate, but also socially and culturally. The well-to-do Darwin family themselves exemplified such choices with several marriages of cousins from a few closely related families of similar religion, social class and social aspirations. Charles Darwin himself made a 'good' marriage to a cousin in the Wedgwood family – they of the world-famous line of pottery – and as a result became financially independent and could devote himself to his science while his

We have long selectively bred plants and animals in order to achieve a desired result.

Ethnic differences between modern humans do not prevent us from making babies because we are all one species and a very young species at that.

Darwin was deeply interested in breeding domestic animals such as pigeons. He hoped that study of successive generations and any changes between them would reveal the mechanism of inheritance but that eluded him. Little did he know that an Austrian monk, Gregor Mendel, had already solved part of the problem with breeding experiments on peas.

Emerging Asian problems

Whilst the genetic evidence for the origin of modern humans in Africa is still overwhelming, there are emerging problems about what has been called Out of Africa I. The concensus view still sees an early dispersal of a *Homo ergaster/erectus* people within and out of Africa around 2 million years ago. The current prevailing view is that they moved northwards into Eurasian localities such as Dmanisi in Georgia by around 1.7 million years ago. But we also now know, thanks to redating of some of the finds, that classic *Homo erectus* was present in Java by 1.8 million years ago.

Furthermore, in 1995 a jaw fragment found at Longgupo in the Central Chinese province Chongqing was claimed to belong to a new species *Homo wushanensis* of considerable ancestry perhaps as old as 1.9 million years ago but this has now been shown to be a fragment of an ape jaw. But continued excavation at the site by a Franco-Chinese team has revealed a horizon containing numerous bovid and cervid bones, that are at least 1.6 million years old. The team currently claim that the beasts were selectively culled by some ancient but, as yet unknown, human relatives and argue that older stone tools found at the site are genuine human artefacts.

Similar claims of great antiquity have been made on tools found at Renzidong (Renzi Cave) in the east of China that have been dated to between 2.6-2.2 million years old. Again doubts have been cast on the reliability of these dates. Nevertheless, more reliably dated stone tools, around 1.7 million years old, of comparable age to those found at Olduvai in Africa, have been found recently at Majuangou in the Nihewan Basin of northern China. In addition there is evidence that the stone tools were used for processing animal tissues and this is the oldest record of such behaviour in east Asia.

The site preserves lake-margin and wetland environments with stone tools being found at four successive horizons stretching from 1.66-1.32 million years ago and are well constrained by magnetostratigraphic polarity data measurements. Abundant tools were found – successively 95, 111, 226, and 443 – nearly 900 altogether, and in places in concentrations of 620 tools per cubic metre of sediment. Associated animal bones include elephant, horse, hyena, rhinoceros, deer, bovid, ostrich and undetermined carnivores. Some of the bones have percussion and scratch marks on them but it is difficult to tell whether these result from trampling by the larger beasts or human

related activity. However, a few of the deer and horse limb bones have been broken open to expose the marrow and it is reasonable to conclude that is the result of intentional human related activity.

Altogether, the mounting information from eastern Asia points towards a quite widespread human related presence in the region by at least 1.7 million years ago. Such antiquity suggests that the move out of Africa was at least 1.8 million years ago. And, if it was by *Homo ergaster/erectus* people, it was almost as soon as they evolved in Africa, unless they belonged to some more ancient species such as *Homo habilis* or *Homo rudolfensis*. Furthermore, this was a move out of Africa before the onset of significant climate oscillation in mid Pleistocene times. Some experts see all this Asian data as a good reason to reassess the Out of Africa I story and point to the curious lack of any fossil evidence of appropriate age en route from Africa to Asia.

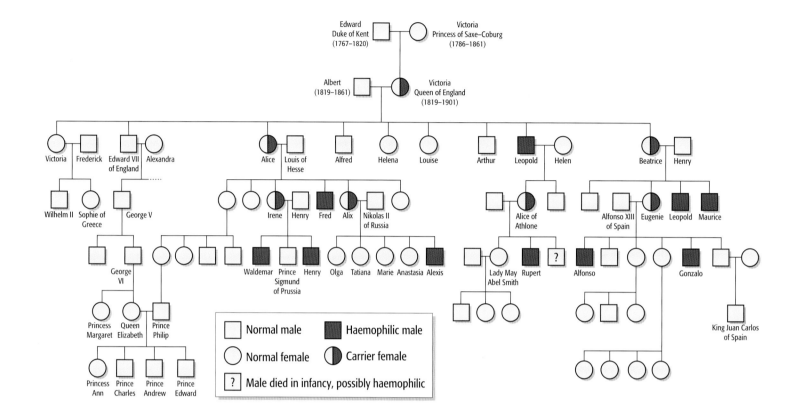

| Normal male | Haemophilic male |
| Normal female | Carrier female |
| ? Male died in infancy, possibly haemophilic |

Human family trees, especially those of royalty who have been obsessed with primageniture and succession, often show up the transmission of inherited traits such as the transmission of potentially lethal haemophilia through successive generations of Queen Victoria's extended family.

investments provided him and his large family with a good income. For many centuries, European royal families have used selective marriage for political, economic and dynastic reasons, as have many different aristocracies around the world – including some famous intellectual 'dynasties' such as the Darwins and Huxleys. Sometimes, however, family and dynastic inbreeding has been so close as to preserve deleterious genetic traits such as haemophilia that recurred over several generations in European royalty since the 19th century.

Genetic measures of difference

Our visual acuity, ability to detect colour and pre-empt the intent of others through reading body language and 'guesstimating' what others are thinking have both benefits and drawbacks. We are very good at detecting difference in others and actively promote similarity through choice of sexual partner and other cultural and social decisions throughout our lives. But we are also capable of incapacitating ourselves with fears and suspicions about other people or even the surrounding environment – and these feelings may have no foundation in reality.

Over the millennia the human diaspora from Africa through the global variety of landscapes and environments has resulted in some populations becoming more isolated than others. The discovery of the extent of human dispersal around the globe and the resulting differences between separate populations fascinated early European explorers. For example, for centuries, European royalty and aristocracies flaunted their wealth and influence by having exotic servants and slaves. In the 19th century Great 'World' or 'Empire' Exhibitions held in European capitals such as London and Paris portrayed such differences with tableaux from far-flung places. These cultural and biological exhibits included live plants, animals and people. Ordinary people were able to see the variety of humanity for themselves and flocked to gaze at this human 'zoo'. Experts examined, described and discussed their intimate details.

Saartjie Baartman (1789-1815) is perhaps the most famous – or infamous – example of this. This Khoisan woman was born in what is now the Eastern Cape Province of South Africa and in 1810 travelled to England, on the promise of becoming wealthy. She travelled around the country showing the locals what they considered to be her very 'unusual' features — namely very large buttocks — which (for an extra fee) they were allowed to touch. Her genitalia were also unusual, but Saartjie did not allow this trait to be exhibited while she was alive. Her exhibition caused a scandal in London and she later moved to Paris, where she was exhibited — by an animal trainer — for 15 months. French anatomist Georges Cuvier was among a number of French scientists who visited her while she was in France.

Saartjie died in 1815 of an inflammatory ailment, but it was only in 2002, after legal battles, that her remains were returned to her birthplace and interred.

The concept of race has developed over centuries and it is now badly tainted with all the disastrous tendencies of destructive racism. Terms such as 'population' are more neutral but not so useful, so 'ethnic group' has arisen as a substitute term for race.

In the 1960s the Italian/American geneticist L. Luca Cavalli-Sforza conducted one of the first attempts to measure, on a global basis, genetic differences between ethnic groups. He measured the genetic difference between 12 loci, mostly coding for blood group antigens, taken from individual representatives of seven major human populations. He found that of the measurable differences less than 15 per cent occurred between the ethnic groups, the remaining and much bigger 85 per cent difference occurred within ethnic groups.

In the 1970s the American geneticist Richard Lewontin carried out a similar but more extensive study. Lewontin divided the global population into seven ethnic groups (he called them races) most of which have been long recognised — namely Black Africans, Mongoloids, Caucasians, South Asian Aborigines, Amerinds, Oceanians and Australian Aborigines. Then each ethnic group was divided into populations; for example the Caucasians were separated into Arabs, Armenians, Basques, Bulgarians, Czechs and so on and recognised by linguistic, cultural, historical and morphological differences. Lewontin then compiled a huge data set derived from nine blood groups, four serum proteins and four other enzymes.

LOVE and BEAUTY — SARTJEE the HOTTENTOT VENUS.

The tragic figure of a Khoisan woman from South Africa who was persuaded to exhibit herself in Europe in the early 1800s on the promise of becoming rich. At the time there was much ill-informed and fundamentally racialist ideology in circulation.

The study again showed that over 85 per cent of the genetic variation found within living humans is contained within individual populations and only 14 per cent or so relates to differences that separate populations. Under half of the 14 per cent relates to differences between ethnic groups.

Our present genetic connectedness

As we have seen, we humans share 98.5 per cent of our genetic material with the chimpanzees and that percentage increases to 99.4 per cent if we only consider those stretches of DNA that contain information for protein synthesis. It certainly seems to be a miniscule percentage difference to account for what we typically perceive as the anatomical, behavioural and cognitive 'gulf' that separates us from the chimpanzees. However, there are many experts who would place the chimpanzees as just another human species within our genus *Homo*, on the grounds that many, many other genera contain species that show just as much biological difference as we do to the chimpanzees.

Nevertheless, this apparently small difference needs to be placed in context. Although the chimpanzee genome is still being mapped, enumerated and analyzed there is an indication that there are regions of the human genome that might not be represented in the genome of the

Is the chimpanzee simply another human species?

chimpanzees or other higher apes and these could be due to either insertions in the genomic lineage of hominins (humans and our extinct relatives over the past five or six million years since the divergence with the chimpanzees) or deletions in the chimpanzee lineage.

By comparison with the mouse lineage, such duplications are to be expected. It is now some 75 million years ago, in late Cretaceous times, since the divergence of the mammal lineage when mice and humans shared a common ancestor. Over that long period several clusters of mouse-specific genes have arisen that are represented only by single genes in the human lineage. Not surprisingly, they are to do with

traits that are particularly important to mice such as olfaction (sense of smell), immunity and reproduction. They indicate the important role that ecology, pathogens and sexual selection play in shaping the differences between mammal groups through their genetic makeup. Nevertheless, it is also important to realise that some 80 per cent of mouse genes have a direct counterpart in the human genome and that more than 99 per cent have some similar counterpart – despite the 75 million years since we last shared a common ancestor.

Geneticists have chosen to search for adaptive genetic changes among proteins and their coding sequences. And, since each protein is some 400

The human cell contains 22 pairs of chromosomes that match in size and shape and a further two, the X and Y sex determining chromosomes, that do not match, making 46 chromosomes in all.

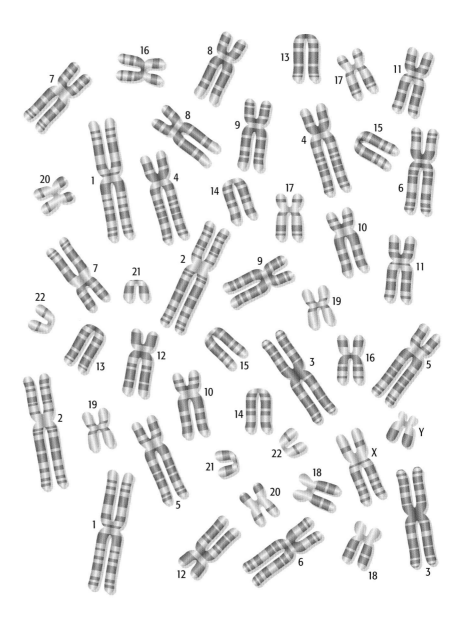

amino-acids in length, it has been estimated that the total number of amino-acid replacements in the human lineage could be as many as 200,000. Of these, less than half are thought to result from positive selection and that means that since the chimpanzee-human divergence there may have been around 70,000 positive adaptive substitutions of amino-acids in the human lineage. If this estimate is in any way correct, the implication is that there may have been one or two adaptive substitutions in every human protein since the chimpanzee-human split five or six million years ago. With such a high number it will be a huge challenge to identify specific changes that are biologically meaningful from the abundance of those that are not.

We would like to know how many genes have made the difference between the apes and us. Is it tens, hundreds or thousands, and which specific genes have made the most significant differences and what were the kinds of changes – changes in regulatory sequences, gene duplications and so on?

Present expert opinion favours changes in regulatory sequences, in other words the way that development is controlled by genes turning on and off at the right time and in the right place.

There have been a few pioneering studies that have looked at the genetic basis for the chimpanzee-human divergence and are beginning to reveal specific genes that might have been implicated in the kind of selective changes that have directed the recent course of human evolution. However, to implicate a specific gene in human evolution there has to be evidence that the gene is functionally involved in the particular trait (that may be physiological, behavioural or developmental) and this may be derived from mutations at a particular locus. Once identified, that locus has to be examined for evidence of natural selection through population genetics and molecular evolution. Recently, a few genes have been identified through their role in human mutations that do clearly affect morphological and behavioural traits.

Over the last 65 million years of Cenozoic time, the Primates have evolved and diversified with several branches dying out only to be replaced by new ones. Even with the benefit of hindsight it is difficult to pinpoint any moment in the past when the future success of the human lineage could be predicted.

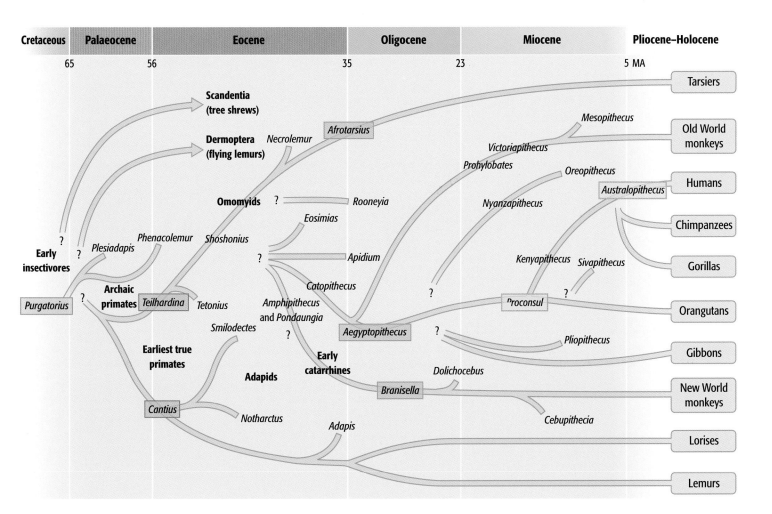

It is likely that there are other genes besides FOXP2 that are responsible for our ability to communicate in speech.

A gene for speech?

Recent developments in medical genetics has allowed the identification of a gene called FOXP2, which is short for forkhead box P2. Mutations of FOXP2 have been implicated in an extremely rare but specific inheritable disorder of speech and language. The gene is not novel to humans but is found in other mammal species. The human FOXP2 protein differs from that found in the gorilla and chimpanzee by only two amino acids and from the orang-utan and mouse by three and four amino acids respectively. Fixation of the two amino acids is estimated to have occurred within the last 200,000 years, coincident with the evolution of *Homo sapiens* and perhaps also coincident with the evolution of language. It therefore suggests that the gene has been a target of positive selection but as yet there is no biological evidence that these amino acid replacements have any functional significance. Interestingly, another implication of this date, if indeed it is accurate, is that it is too late for the Neanderthals to have benefited from the same all-important innovation of language, unless it arose independently in them – which is unlikely.

The FOXP2 gene is a kind of 'master switch' that encodes a protein promoting the expression of other genes and directs an aspect of development. This can be seen in human foetuses with FOXP2 expressed in specific brain loci that are destined for fine motor control such as the caudate nucleus and cerebellum. Indeed brain scans of people suffering with defective FOXP2 show these regions to be abnormally small. It is possible that the unique human version allows the specialised and synchronous tongue and mouth movements that are necessary for speech. However, it must be remembered that the FOXP2 gene is unlikely to be the only gene involved in the development of language.

Genes for brain expansion and jaw contraction?

We know that the development and expansion of the brain has been an ongoing but not steadily progressive process over the last six million years. There have been long periods of very slow growth, especially early on when successive australopithecines showed very little brain expansion. Then around two million years ago, the rate of increase accelerated with species such as *Homo rudolfensis*.

Our body's 'building code'

Genes encode all the information necessary for the construction and running of any organism whether microbe or man. Genes are made of deoxyribonucleic acid (DNA for short) in the form of two long helically coiled spiral strands that form the chromosomes. Each of the two coiled chromosome strands is made of sequences of four types of small molecules: adenine, guanine, cytosine, and thymine (labelled A, G, C and T for short). Their sequence pattern encodes and carries all the genetic information that is transferred from parent to offspring in reproduction. Forty-six pairs of chromosomes are found in every cell nucleus in the human body and the DNA they contain is known as nuclear DNA (nDNA) and each human body has some 100 trillion cells. The nDNA of the Y-chromosome is particularly useful as it determines male sexuality in humans and therefore can be used to track male ancestors over time.

But genes only make up a very small part of all this nDNA, only about 10 per cent, in fact. The rest is often referred to as 'junk DNA' and consists of repetitive DNA sequences, both long and short, with the latter being referred to as 'microsatellites'. Also, there are 'pseudogenes', which are segments of DNA that were genes in the past but have been inactivated or 'turned off', plus 'introns' which are segments inserted in the middle of genes. However, since this 'junk DNA' also mutates over time and is copied along with encoding DNA, it can still provide important information about evolutionary relationships.

DNA is also found outside the nucleus but still within the cytoplasm of the cell. It is contained within numerous small mitochondria, which provide the energy for the body's metabolism. This mitochondrial DNA (mtDNA)

Mammoth DNA

The recovery and analysis of ancient DNA is fraught with problems and although the technique holds out great promise it is very difficult to find suitable material that is more than a few thousand years old. But luckily there are dedicated and persistent teams of molecular biologists who have refined meticulous techniques that can obtain replicable results. Most recently some ancient mammoth and Neanderthal DNA has been recovered and is providing interesting new information.

Of all the animals of the Ice Ages, the woolly mammoth, *Mammuthus primigenius*, holds out one of the best hopes for the recovery of ancient DNA because so many bones and even flesh has been preserved in subzero, permafrost conditions since many of the individual animals died tens of thousands of years ago. Very fragmentary mammoth DNA has previously been recovered but not in sufficient quantity to finally determine whether or not they were more closely related to the living African or Asian elephants which are placed in separate genera – *Loxodonta africanus* and *Elephas maximus* respectively. There is fossil evidence that this whole group of 'elephants' (proboscideans) diverged in Africa around 6 million years ago but the relative timing of the mammoth divergence was not clear.

Now, we know that the mammoth is more closely related to the Asian elephant but only just thanks to new samples and new techniques. Well preserved mammoth bone from the Arctic Russian mammoth 'graveyard' of Berelekh and new amplification techniques have allowed analysis of the complete mitochondrial genome of a mammoth.

The new analysis indicates that the African elephant lineage split off first followed by the separation between the mammoths and Asian elephants some 440,000 years later.

The Berelekh mammoth DNA is only some 13,000 years old but now some 100,000 year old Neanderthal DNA has been recovered from the tooth of a young Neanderthal girl who lived and died in the Meuse Valley (Sciadina Cave) in what is today Belgium. Previously, the oldest Neanderthal DNA has been around 50,000 years old and when compared with some 28,000 year old Neanderthal DNA seemed to suggest a fairly consistent Neanderthal gene pool in Eurasia over a significant period. The new data shows some greater divergences from this gene pool and from that of modern humans.

is effectively inherited solely from our mothers and so traces female descent and development over time. As it changes by mutation much faster than nDNA, the variation in mtDNA provides important information about the timing of speciation as well as identifying evolutionary relationships. The assumption that both nDNA and mtDNA have a constant rate of mutation provides the basis for the so-called 'molecular clock' (see box).

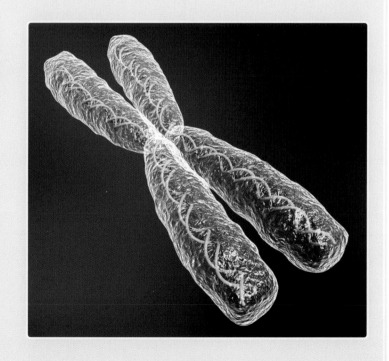

Our 'molecular clock'

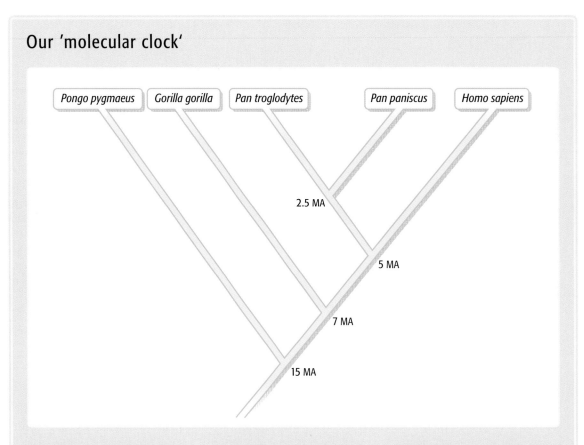

The 'mechanism' of the so-called 'molecular clock' is driven by the assumption that the rate of mutation in mtDNA is constant over time. So, by counting the number of unique and shared bases along any strand of mtDNA within a given population it is possible to then calculate the 'molecular distance' between populations. Equally, a measured molecular distance between species should be proportional to their separation in time since they last shared a common ancestor. However, it is important to realise that since DNA becomes differentiated before populations, the measures will not be exactly the same.

For example, within the apes the greatest mtDNA distance (5 per cent) is between the gibbons and the great apes (orang-utans, gorillas and chimpanzees) and humans, indicating that this is the earliest divergence. This was followed in time by the orang-utan, which has a 3.6 per cent mtDNA distance from the remaining gorillas, chimpanzees and humans. The gorilla has a 2.6 per cent mtDNA distance from the chimpanzees and humans, and the chimpanzee has a 1.6 per cent mtDNA distance from humans. The common and bonobo chimpanzee species only differ from one another by 0.7 per cent. Since the measured chimpanzee-

human distance is approximately half that of the orang-utan-chimpanzee distance, and the fossil evidence shows that the orang-utans diverged from other apes somewhere between 12 and 16 million years ago, it then follows that, according to the molecular clock, the chimpanzee-human split occurred somewhere between 4.2 and 6.2 million years ago, the gorillas diverged between 6.2 and 8.4 million years ago, and the gibbons some 18 million years ago.

Recently, a mutation of a gene called ASPM has been found that makes the brain abnormally small, suggesting that the gene controls brain size. As with FOXP2, there is a suggestion that the gene has been positively selected for over a long time, perhaps even before the chimpanzee lineage diverged from the human one, because it performs a similar function in the great apes. The biological function of the gene may be the control of cell division in the developing brain but there is a limiting factor. The size of the ape cranium — the bony brain case, is limited by their need for a powerful jaw musculature to cope with their plant food – again, a typical evolutionary compromise. For the cranium to 'balloon' upwards to accommodate a larger brain the jaw musculature and the jaw itself had to diminish in size, and that could only have happened with a change of diet to food that can be more easily pre-processed than plant fibres – meat.

Another gene called MYH16 is a critical muscle protein that is present in nonhuman primates but missing in humans. It is thought that the gene was 'knocked out' (suppressed) some two million years ago just when early human species such as *Homo rudolfensis* and *Homo ergaster/erectus* evolved.

By the 1980s, technological advances had made the complex business of DNA analysis much faster and cheaper. The first study of variation in the human population using mtDNA was conducted by Wesley Brown and was based on a small sample of 21 people. What few differences that could be detected were explicable by mutations accumulated since the divergence of the human population from its ancestral population. Since the rate of change (of nucleotide substitutions) in human mtDNA is estimated to be 1 per cent per million years and the detected variants differed from the ancestral sequence by just 0.18 per cent, Brown calculated that they had taken just 180,000 years to accumulate. His human samples could be divided into three ethnic groups, for which the measured variants showed that the Afro-American population diverged earliest, followed by the Mongolian and then the Caucasian group.

The 'lucky mother' – mitochondrial 'Eve'

A much more famous study in the late 1980s caught public imagination around the world through being nicknamed the 'African Eve' or 'lucky mother' hypothesis by the media. A Californian team comprising Rebecca L. Cann, Mark Stoneking and Allan C. Wilson analysed mtDNA from 147 individuals drawn from five global populations – African, Asian, Caucasian, Australian and New Guinean.

They surveyed 467 sites in each of the individual samples thus encompassing about 11 per cent of the 16,569 nucleotide pairs that make up the human mtDNA molecule. Nearly half the sites showed variation, revealing 134 types of mtDNA that

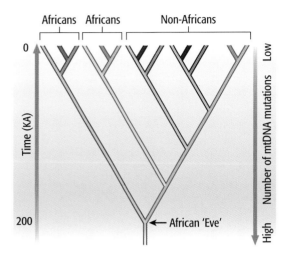

seemed to fall into two distinct branches, one containing African individuals while the other included the other groups and some Africans.

The authors came to the same basic conclusion as the earlier study by Brown, namely that all the mtDNA diversity originated from a single African woman who lived around 200,000 years ago and that all the populations, apart from the African one, had multiple origins suggesting that each area was colonised repeatedly since then.

The results were published in the international science journal *Nature* (1987, volume 325, p.31) and the accompanying editorial began with the remark that 'A paper by R.L. Cann, M. Stoneking, and A.C. Wilson...reports that Eve was alive, well and probably living in Africa around 200,000 years ago'. The media promptly picked up that bit of journalism and soon 'African Eve' became a global marketing phenomenon and there was no stopping her.

All the previous studies that had come to similar conclusions were ignored, and yet the Eve metaphor is more than somewhat misleading. It must be borne in mind, however, that 'Eve' was not a lone female, but one of several thousand who formed part of that ancestral population 200,000 years ago.

The results came under intense scientific scrutiny and soon began to show cracks. One of the problems was that the way the authors had constructed their tree was found to be unreliable; it turned out not to be the most 'parsimonious' (on the grounds that 'nature' tends to be 'economical' in other words, does not use anything more than the absolute minimum) as the separation of the Africans from the rest was not as clear as claimed.

Amongst modern human populations around the world, the greatest variation is found in Africa. The implication is that populations within Africa must have been the first to diverge and allow time for those variations to evolve. Consequently, all modern humans must have developed from an original African population.

Technological advances over the past 25 years or so now mean DNA analysis is faster and cheaper than in the past.

Furthermore the 'Africans' turned out to be Americans of Afro-Caribbean descent and the estimated timing of the divergence had a large standard error and could have occurred as long as 800,000 years ago. Luckily, follow-up studies increased the sample size and the database, and included chimpanzee DNA to root the tree more securely, but the general conclusions were just the same as the earlier studies had shown, with a refined divergence time of 238,000 years ago.

The Darwins' mtDNA

Just as Charles Darwin and his five siblings inherited their mtDNA from their mother, Susannah Darwin (née Wedgwood), and all of Darwin's 10 children inherited their mtDNA from their mother, Emma (née Wedgwood – Emma was Charles' first cousin), none of Darwin's grandchildren inherited Emma's mtDNA because all were born to Darwin's sons and none to his daughters. However, there is a woman alive today whose mtDNA will be ancestral to all mtDNA in the distant future. That is why the nickname 'lucky mother' is more appropriate – but it has not caught on in the same way.

Genetic analysis of living dog populations show that they all have evolved from a wolf ancestor that lived around 6,000 years ago.

In an attempt to further reduce the margin of error in the divergence time, a Japanese team led by Satoshi Horai sequenced the complete mtDNA genome from three humans, an African, Asian and European, as well as that of four great apes, a common chimpanzee and bonobo, a gorilla and an orang-utan. Taking the divergence time of the orang-utan from the African great ape plus human lineage at 13 million years ago, they estimated that the human-chimpanzee divergence occurred 4.9 million years ago and that the divergence between African and non-African members of our species happened 143,000 years ago – give or take 18,000 years.

Nuclear DNA backs the 'Out of Africa' hypothesis

This kind of DNA encodes most of our body structures and we inherit it from both parents, but it also contains great chunks of so-called 'junk DNA' (see box on p. 174) that nevertheless mutate and can give us information on evolutionary relationships. Recently, an analysis of a particular locus on chromosome 12 indicates that while African populations vary considerably with many patterns, non-Africans have only one pattern. Yet again, it seems that all non-African populations have descended from an African one that probably occupied north-east Africa around 90,000 years ago.

What about the Y's?

In terms of transmission, the Y chromosome is the male counterpart of the mtDNA and mostly does not recombine. Consequently, it can serve as a travelogue through time, which records the accumulation of small changes (called polymorphisms) in the chromosome. Unlike mtDNA it does not vary a great deal because it has a much smaller population and is transmitted from father to son. The smaller population is due to females carrying two X chromosomes but no Y, while males carry one X and one Y. When combined in sexual reproduction the ratio is three X to one Y, hence the much lower population of Y.

In the late 1990s an American team led by Michael Hammer sampled 1544 males from 35 populations at eight sites within a region of the Y chromosome. The alleles (alternative forms of a specific gene) at the eight sites occurred in 10 combinations or haplotypes. Five of the haplotypes are restricted to a single continent – three to Africa, one to Asia and one to Australia. Africans and Europeans share two haplotypes, Africans, Europeans and Asians share one, and two are present in all regions.

All in all, eight haplotypes are present in Africa, five in Europe, four in Asia, three in Australia/Oceania and two in the New World. When the data are formed into a network it becomes evident that the ancestral haplotype is restricted to Africa and has a low frequency of occurrence as would be expected if Africa were the base from which all subsequent differentiation originated. African populations south of the Sahara have the highest number of haplotypes restricted to a single continent and the highest number overall (eight), which is to be expected as the oldest Y-chromosome population with the longest time to accumulate the most mutations.

The team also used their data to estimate the timing of the origination of the nine haplotypes from the ancestral one as around 147,000 years ago with a standard error of 51,000. And, the timing of subsequent mutations ranges from 110,000 to 1,000 years ago. It was the 110,000 mutation that became the most common and widespread Y-chromosome haplotype and reflects the diaspora of *Homo sapiens* beyond Africa. This genetically based chronological record is still carried around by most men and its timing is in good congruity with the record of bones and stones left behind by our African ancestors.

Altogether the evidence for our African ancestry is imposing, as is the timing of the initial dispersal of modern humans beyond Africa at around 100,000 years ago. Molecular evidence can also be combined with other methods to address more detailed aspects of this dispersal, such as the question of the occupation of the Americas.

New populations, new languages

There are an estimated 6912 living languages in the world today although many are in danger of dying out. They range from Aari, spoken in northern Ethiopia, to Zyphe, a language of Myanmar. Africa alone has over 1000 and Australia around 600. These languages and language groups seem to be straightforwardly related to specific populations and cultures. Consequently there has been an assumption that the division and diversification of the world's languages reflect the development and splintering of a growing human population from a small founder population with language groups and gene pools sharing a common structure.

The idea that languages and cultures should represent genetic populations is based on the observation that languages, like genes, are passed from parents to children and 'marriage' of the parents tends to be constrained within the same language group.

Therefore the global number of languages should be telling us about the history and antiquity of the human population. Also, if the historic linkage between different languages could be established and traced geographically it would provide an independent test of the ancient movements and divisions of humanity. Furthermore, with this model in mind, within all languages there should perhaps be some common elements linking them to the original one from which they were all derived, the so-called 'ur' language. Despite many attempts to establish what that language might have been like, it has so far proved elusive.

That Africa should have so many languages comes as no surprise since it supports all the other evidence for an African origin of all modern humans. And equally, the apparently surprising

Polymorphisms have been discovered only in the past few years, and even today relatively few have been described.

Comparison of language and gene trees of relatedness show significant concordance and reinforce the fossil evidence for the dispersal and separation of different ethnic groups of modern humans over the last 100,000 years.

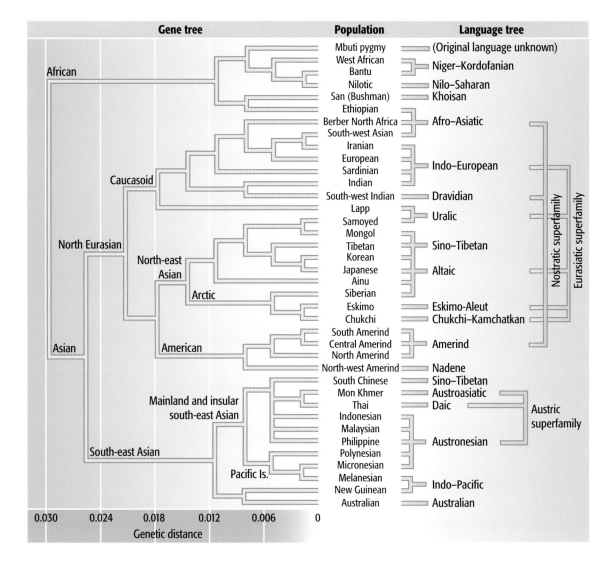

number of Australian languages just reflects the early occupation of that large continent by a founder population. The process is thought to go like this:

A small 'founder' population enters a new territory and as that population grows and threatens to over-use local resources, pioneer splinter groups break away from the parent group and move into new territories, which are geographically separate. Isolation then helps promote cultural, linguistic and genetic changes within the new populations. Consequently there develops a tree-like structure from the ancestral stem to the descendent branches reflected in genes, culture and language. The greater the distance in both space and time between populations, the greater the differences should be. At least this is the story we tell ourselves.

With modern techniques of genetic analysis and well-established linguistic research and analysis of the world's wonderful diversity of languages we should be able to test the veracity of this hypothesis. Indeed, there have already been a number of tests 'run' in different parts of the world, but they have produced mixed positive and negative results and no general synthesis because of the several different methods of sampling and comparison have been tried – and the experts have yet to agree on which is the most appropriate.

A new study has tested the idea in North America. The development of languages in this continent is of particular interest because of North America's relatively isolated geography and history of human occupation from Asia.

Like Australia, North America has only one viable geographical entry point and a restricted one at that. It's the Bering Straits between Asia and North America – a narrow and shallow seaway, which has been exposed as a land bridge whenever sea levels have been lowered during past glaciations.

At present the archaeological and genetic evidence points to the entry into North America from Siberia of a single Asian founder population at least 15,000 years ago and most likely considerably earlier – perhaps even 40,000 years ago. The Bering land bridge, known as Beringia, acted as a bottleneck and temporary 'freeway' for both migrant animals, such as mammoths and horses, as well as modern humans. We know that these human migrants spread rapidly south and crossed into South America to reach southern Chile by around 14,500 years ago. Because of this widespread geographical dispersal across several climatic and geographical barriers it is highly likely that when stable populations grew from the founder groups they would have some significant cultural and linguistic and some perhaps detectable genetic differences.

When Europeans first 'rediscovered' the Americas, there was a rich diversity of languages in the continent. And when the first investigations of possible links between linguistic and genetic groups were investigated it did indeed seem to verify the expectation that the average genetic distance between populations within different language families were greater than average genetic distances between populations within language families.

However, when the mtDNA from 17 native North American populations (ranging from the Aleuts to the Cherokee) was compared with language groupings in this context the idea does not work – or rather the 'hypothesis was rejected', to use more correct language. Furthermore, the genetic structure frequently departs from the linguistic structure that has been mapped out, showing that there has been significant gene flow across the linguistic boundaries.

This shows that, in North America at least, the language of a population can be replaced more rapidly than its genes because language can be transmitted both vertically from parents to children and horizontally to unrelated people while genes cannot be transferred 'horizontally' to unrelated people. Paradoxically however, there are often strong cultural pressures, which maintain a language while genes may be 'freer' in their flow across language and cultural barriers in certain situations, especially in our crowded modern world with wide-ranging and rapid travel.

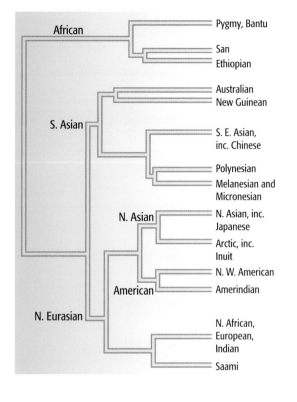

Movements of modern peoples over the last few hundred years has resulted in many people inhabiting regions other than those of their ethnic origins. For instance, present American populations include people of European and Afro-Caribbean origin as well as native Americans, who themselves originated from Asia. English populations include people of Celtic origin as well as those from many other locations around the world.

We might look superficially different but it is only skin deep, genetically we are one species. And whilst linguistic and cultural differences can make the differences between life and death in conflict situations, they can be overcome in a few generations.

Often, strong cultural pressures maintain a language while genes may be 'freer' in their flow across language and cultural barriers.

How old is it? When did it happen? ...
Dating the history of humankind through time

Up to the mid-20th century it was possible to date archaeological and palaeoanthropological materials and specimens only by relative methods (see box on stratigraphy and fossils – page 86). But after World War II and developments in technology relating to detection and measurement of radionucleides, it became possible to use radiometric methods to measure the time that had elapsed since the formation of certain kinds of naturally occurring materials. These ranged from organic materials such as animal and plant carbon-based tissues such as bone and wood (radiocarbon dating) to the formation of certain minerals in inorganic rocks such as lava and volcanic ash.

Eminent scientists such as the New Zealand-born physicist Ernest Rutherford (1871-1934) and the British chemist Frederick Soddy (1877-1937) established the principles of radiometric dating early in the 20th century. In the early 1900s they discovered that it was possible to predict rates of radioactive decay for certain chemical elements and they suggested that the gas helium might be a product of such decay.

From the moment they are first formed, certain common radioactive elements decay at known constant rates from the 'parent' through one or more 'daughter' isotopes. By measuring the ratio of the 'parent' to 'daughter' isotopes remaining and knowing the rate of decay, the time of formation can be calculated.

But it was the American radiochemist Bertram Boltwood (1879-1927) who first determined an understanding of the uranium decay series and developed the analytical techniques for measuring the results. Boltwood obtained dates from certain rocks containing radioactive minerals, including one from Ceylon (today's Sri Lanka) that was 2200 million years old. In doing so he increased the predicted age of the Earth by an order of magnitude.

By 1910, Arthur Holmes (1890-1965), a British geologist, had arranged a number of such dates in chronological order against estimated thicknesses of rock strata belonging to the succession of periods of geological history. The result was the construction of the first geological time-scale. As more dates became available he refined the scale, progressively creating a more and more complete record.

By 1937 the oldest fossiliferous rocks were known to be over 450 million years old and strata which bore primate fossils such as the Miocene began around 32 million years ago (now known to be 23 million years old). The base of the younger Pliocene seemed to date from around 13 million years ago (now known to be 5 million years old) and we now know that the base of the Pleistocene is about 1.8 million years old.

The problem with this kind of radiometric dating is that of relating it to palaeoanthropological material unfortunately it is not possible to use it to date organic materials such as bone directly, and there is no point measuring the age of the stone from which a tool is made because the stone's age has essentially nothing to do with the date at which it was selected and modified into a tool by a human or human-related hand.

However, the method is very important as an indirect dating method. It is particularly useful where radiometrically datable rocks such as volcanic lavas and ashes were commonly erupted in environments where our ancient relatives lived.

Fortunately, the Great East African Rift Valley and the islands of Indonesia have been active volcanic regions for several million years. Their numerous highly explosive volcanoes have ejected large amounts of volcanic magmatic material, in the form of lava and ash, over the surrounding landscapes.

These volcanic rocks often contain minerals and radioactive isotopes (potassium 40, which decays to the gas argon, is one such isotope) that can be radiometrically dated. The rocks also become interbedded with sediments and also the potential fossils remains of our early human relatives, their tools and the remains of the animals and plants that occupied the same environment as them. Therefore most radiometric dates refer to rocks that lie relatively close to the fossils that we want to date. However, even when the method is as accurate as possible it still has a quantifiable margin of error that generally amounts to thousands of years for specimens that are more than a million years old.

For instance, in the famous Olduvai Gorge section of Tanzania excavated by the Leakeys, the occurrence of lava and layers of volcanic ash provided the possibility of dating the strata, and therefore the fossils and artefacts embedded in them.

Potassium-argon dating of the lava at the bottom of the exposed succession gave an age of nearly 1.9 million years. This was very close to the base of the Pleistocene and provided an important link to a period of normal polarity in the Earth's magnetic field known as the 'Olduvai event' (see box on palaeomagnetism, page 120). So the discovery of fossils that included Homo habilis and crude stone tools near the base of the succession could be given a maximum age of 1.9 million years.

Unfortunately, the problem with carbon isotopes is that their rate of decay, known as their half-life is relatively rapid and normally there is not enough of the parent material left to measure after 40,000 years. But with the development of accelerator mass spectroscopy (AMS) it became possible to measure much smaller amounts and the method can now be used for material up to 80,000 years old. Even this, however, is still is too young for most palaeoanthropological material such as the bones of our extinct human relatives.

Another problem with radiocarbon dating that has been realised in recent years is that the incorporation of radioactive carbon into the tissues of animals and plants is more complex than previously thought so that there is no simple linear relationship.

The result is that there is a measurable difference between radiocarbon and calendar years, which can amount to over 1000 years on dates that are tens of thousands of years old. Scientists are presently working to calibrate the discrepancies.

In the last few decades, fortunately, a number of new radiogenic dating methods - for example thermoluminescence (TL), electron spin resonance (ESR) and fission-track dating - have become available and can be particularly useful in certain settings and with particular materials. They basically depend upon generation of ionising radiation that propagates into the surrounding environment. For instance, TL dating uses the decay of radioisotopes that release energy in the form of displaced electrons, which become stored within the crystal lattices of adjacent minerals and accumulate over time.

When such minerals are subsequently sampled and appropriately stimulated (heated) the electrons are freed and emit their stored energy as light (luminescence). The intensity of the luminescence reflects the original dose. It can then be measured and calibrated as an indication of the time accumulated since the mineral began to receive the ionising radiation.

From the archaeological point of view this method is particularly useful for dating pottery, which accumulates radiation since it was first fired. In our prehistoric context (before pottery was made) certain sediments can be dated using this method. ESR is another radiogenic dating method that measures the effects of natural irradiation on materials such as bone, teeth, sediment, shells, cave deposits (travertine, stalactites etc) and burnt rock material (flint and so on). The method is limited to materials accumulated over the last two million or so years and has an error margin in the order of hundreds of years. The technique is often used in the absence of any other dating method and as a check on other methods.

Fission-track dating uses the fission of radioactive Uranium 238, which occurs both naturally and spontaneously. The fissile products have enough energy to leave physical traces (tracks) of their trajectories through surrounding materials, especially mineral crystals. By counting the number of fission tracks, knowing the rate of fissile decay and the remaining uranium content of the source it is possible to calculate the accumulated time over which the activity has occurred. The method is particularly useful for dating volcanic products in which the radioactive material was formed and leaves its tracks within rock materials such as obsidian (volcanic glass) and in zircon crystals.

There are a number of other dating methods such as tree-rings, lake varves (glacial lake sediment layers) and amino-acid racemisation but they are more appropriate to younger archaeological materials than the more ancient deposits we are concerned with here.

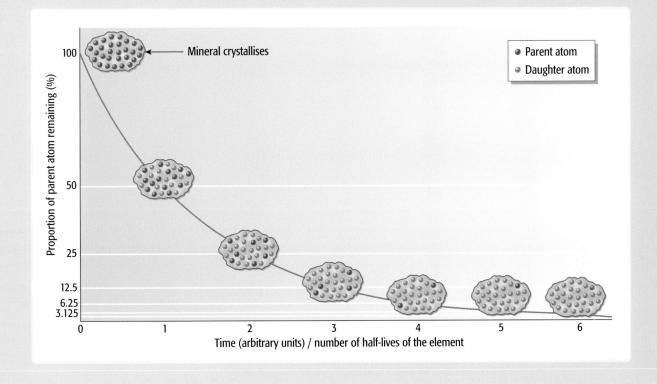

Picturing the human 'family tree' today

As we approach the 150th anniversary (in 2008) of the Darwin/Wallace outline theory of evolution, it is worth finally reviewing how the picture of our family 'tree' looks today. Although the 'tree' icon and metaphor has become intimately linked with the traditional representation of evolutionary descent, the image is much more like the one that Darwin himself originally pictured in his famous notebook sketch of 1837 and then reproduced in his 1859 book on *On the Origin of the Species by Means of Natural Selection*. The image is very shrubby with lots of branching nodes and short twigs from which there has been no further issue.

The fossil record shows just how right Darwin was with this kind of depiction. The history of life in general follows this kind of pattern at all levels, from large groupings such as phyla (the molluscs and echinoderms, for instance) down to the finer kind of divisions that we are concerned with at the family level (in the taxonomic sense).

At present our little, human related bunch of some 20 species scattered over the last seven million years seems to have a narrow base around seven million years ago represented by just one species *Sahelanthropus tchadensis*. However, this is a very recently discovered species and its solitary position is almost certainly an artefact of the poverty of the fossil record at the beginning of Pliocene times (seven to four million years ago). If we look a bit further back into Miocene times, there is a much better record of fossil apes in both Africa and in parts of Asia and Europe.

We now know that the interval from six to four million years ago has just a scattered fossil record in Africa that consists of species belonging to three genera (*Orrorin tugenensis*, *Ardipithecus ramidus kadabba* and *Australopithecus anamensis*), none of which are represented by anything like complete skeletons but rather just a few bones and no skulls as yet.

Moreover, as we have seen, interpretation of these remains is problematic with various disputed claims about whether they were capable of upright walking and how they relate to subsequent evolutionary developments. At the moment there is no consensus on which of these extinct relatives was actually ancestral to our genus. What they all show is a mosaic of primitive and more advanced characters in

> **Opinions are divided as to exactly which of the extinct relatives was actually ancestral to our genus.**

different combinations, suggesting that there was considerable diversity and presumably adaptation to different ways of life at the time with no simple domination by any one species. This is a pattern that was repeated until very recently when our lot, *Homo sapiens*, became the dominant human species.

Until recently it was thought that late Pliocene times from 3.5 to around 2.3 million years ago was solely dominated by members of the small bipedal but still small-brained and ape-like genus *Australopithecus*. The genus includes some very robustly skulled plant-eating species that have always been regarded as an evolutionary cul-de-sac, although they were very successful at the time. Some of them may even have 'invented' the use of stone tools and because of the overall success of the genus, it was assumed that later human evolution had to stem from within this group and the mantle of putative ancestor was placed fairly and squarely on little 'Lucy's' shoulders (*Australopithecus afarensis*). However, the discovery and description of a new genus and species – *Kenyanthropus platyops* ('flat face') has threatened to dethrone 'Lucy' as the skull of 'flat face' shows some distinctly more human-looking features. Since this species dates from around 3.5 million years ago and is thus contemporaneous with 'Lucy', it now seems possible that evolution bypassed the australopithecines all together.

Meave Leakey and her collaborators think that, in retrospect, we can discern an evolutionary line of development towards *Homo* from *Kenyanthropus platyops* to another species *rudolfensis*, which they also place in the genus *Platyops*, and then to *Homo ergaster*, and so on. Needless to say, there is considerable disagreement about this and many experts still prefer the idea that *Homo habilis* was the first species of the genus and evolved from an australopithecine such as *Australopithecus africanus*.

However, there is rather more agreement that from the beginning of Pleistocene times around two million years ago, there was a very successful and long-lived human species called *Homo erectus* that originated in Africa, probably from the closely related species *Homo ergaster*. It was *Homo erectus* that first dispersed beyond Africa into Eurasia and on into southeast Asia where it may have spawned the newly discovered diminutive species *Homo floresiensis*.

The subsequent pattern of evolution in *Homo*

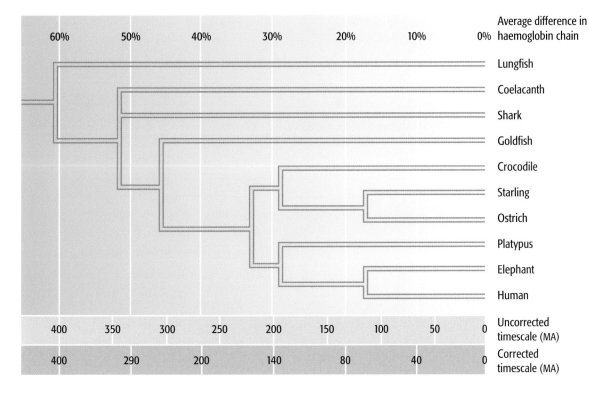

Average difference in haemoglobin chain							
60%	50%	40%	30%	20%	10%		0%

Lungfish

Coelacanth

Shark

Goldfish

Crocodile

Starling

Ostrich

Platypus

Elephant

Human

Uncorrected timescale (MA)							
400	350	300	250	200	150	100	50	0

Corrected timescale (MA)						
400	290	200	140	80	40	0

The evolution of life on Earth a wondrous phenomenon. Over the last 200 years and more tens of thousands of scientists have devoted their lives to the study of living organisms and how they have arrived at their present abundance, diversity and complexity. Every avenue that has been explored from anatomy and general biology to genetics (and here relatedness through measures of difference in haemoglobin is illustrated) and the fossil record independently point to an evolutionary relatedness with a deep history stretching back over hundreds of millions of years.

becomes increasingly debatable with a number of species recognised from Africa and Europe, but whose inter-relationships are far from clear. The general reader can be forgiven for being puzzled by how species can apparently come and go, but we have to remember that a fossil species is not the same as a biological one. A fossil species is purely the construct of palaeontologists who give a name to a group of fossils, which they think belonged to an interbreeding succession of populations over a particular time interval. As we have seen, this is a very contentious area with lots of disputes between experts and new discoveries constantly change assessments of species and genera.

There is one particularly interesting and problematic species in our genus – *Homo heidelbergensis* – which was originally described from Germany but has now had some African fossils assigned to it and lived from around 500,000 and 250,000 years ago. From the temporal and chronological point of view the species is well placed as a possible link between African populations of *Homo ergaster* that lived around 1.5 million years ago and early members of *Homo sapiens* who lived in Africa over 200,000 years ago. However, there is still a significant gap of around a million years between the last record of *Homo ergaster* and the first record of *Homo antecessor* from around 800,000 years ago that fills the gap but

there is not a huge amount of support for this yet. There clearly is a desperate need for more fossils from the interval in time.

At present both fossil and genetic evidence strongly supports the theory that modern humans all evolved from an African population of *Homo heidelbergensis*-like people around 200,000 years ago. But it was not until around a 100,000 years ago that a small founding population of them moved north out of Africa into the Middle East. From this genetically similar population all modern humans have evolved. The one other famous human relative not so far mentioned in this summary is *Homo neanderthalensis*, another cul-de-sac species who is probably derived from a European population of *Homo heidelbergensis* around 350,000 years ago.

It is only with the benefit of hindsight that we can construct this sort of family 'bush' and theorise about possible evolutionary links within it. I do not think that at any time within this evolutionary history it would have been possible to predict the possible outcome that we have today. Equally, I do not think it is possible to predict the distant future for our species. One thing is certain; no species stays the same indefinitely, all have either died out or diverged into new species. But then no species has been able to manipulate it own destiny in the way that we can.

New discoveries are a constant source of debate between experts.

So what does make us human? – The latest developments.

As we have seen, the fossil record of bones and stones convincingly shows us that over the last seven million years our extended family of some 20 or more related species has diverged and diversified into a variety of more or less human-like forms. Most of these fossil relatives differ in some significant way or other from our nearest living relatives the chimpanzees. For instance, most acquired the ability to walk upright and many were no longer just plant eaters but included flesh as part of their diet. As a result their body form changed to a more human-like condition, especially within our close family (members of the genus *Homo*). And, associated with these dietary changes, there was an increasing use and development in the manufacture of tools.

However, the most important evolutionary changes that differentiate the human family from our higher ape 'cousins' are associated with the brain and our cognitive superiority. Curiously, it turns out that despite intensive investigation over more than 150 years, the anatomy and structure of our brains appears to be remarkably similar to that of the chimps. The most evident difference is one of scale. Over the last two million years or so the human skull has expanded upwards to accommodate a threefold increase in brain size over that of our more ape-like ancestors. Clearly size matters but the evolutionary expansion of the brain must have been accompanied by developments at the cellular level that contribute to our cognitive differences. As these changes are heritable they must have left some signature on the human genome.

The sequencing of the human genome and more recent sequencing of the chimp genome has for the first time made it possible to search for the critical genomic innovations that have made the all important cognitive differences. But so far they have proved strangely elusive and very little evidence has emerged that directly links differences in DNA to anatomical and behavioural differences between chimps and humans. So what is going on?

The assumption has always been that adaptive change would be found within DNA sequences that code for proteins. But comparative searches of such sequences between the chimp and human genomes have not revealed any highly significant differences. However, it has now been realised that a significant proportion of the functional DNA sequence does not lie within the protein-coding sequence but rather outside of it within the vast non-coding proportion (around 98.5%) of the genome, the so-called 'dark matter'.

Recently much effort has gone into scanning the known mammalian genomes for non-coding genomic regions that have remained largely unchanged until around seven million years ago when the chimp/human lineages split. Now all this effort has paid off and revealed a human accelerated region (HAR) gene that may well be linked to the acquisition of humanness. It appears that HAR1F has accrued 18 changes over the last seven million years when the normal rate of substitution would be one or two over the same period.

This RNA gene, called HAR1F, is active in early embryonic development when many of the nerve cells of the neocortex region are establishing their role within the developing brain. Many of the our brain's most sophisticated processes are sited in the neocortex. Interestingly, HAR1F is also expressed in the adult human ovary and testis and sexual selection may also be involved in what is probably a complex process of recombination in which an offspring obtains a blend of parental genes.

Although HARs occur in non-coding regions of the genome, they are often found near protein-coding gene sequences that have neuro-developmental functions, especially those that determine when and where these genes are turned on. Thus the rapid evolution of the human brain and related behaviour may thus be more associated with small changes in the position and timing of the expression of protein-coding genes. The search continues for more HARs in the functional but non-coding 'dark matter' of the human genome continues. But is only thanks to the availability of the chimp genome that we have any hope of making sense of any such differences. Unfortunately, it is not possible to determine exactly when or in which species of *Homo* the HAR really took off. Was the process gradual or did it suddenly speed up?

Breaking news – Lucy's 'daughter' found in Ethiopia
The discovery of the skeleton of a three year old australopithecine girl promises to be one of the great finds of the decade. Juvenile fossil remains of our extinct human relatives are exceedingly rare and this, three million year old one, is by far the oldest known.

Unusually well preserved and complete (about 50%), the skeleton belongs to *Australopithecus afarensis*, the same species as the 3.2 million year old 'Lucy'. The find was made in 2000 just 10 km away in the same conflict ridden region of Ethiopia, known as Dikika, but it has taken six years for it to be removed from its hard rock matrix.

Now, for the first time we gain some idea of the early development of these extinct relatives and how they might have differed from the apes and from modern humans. Hopefully this will allow us to home in on those crucial anatomical changes that finally made the difference between being more ape-like and more human-like. The child's skull shows key features that are still quite ape-like and diagnostic of her species. Her brain size is ape-like but its development seems to have been slightly slower than that of the living apes. Interestingly, it is, in this sense, tending towards a more human rate of development. Age is determined by the stage of tooth development and it is the presence of a full set of milk teeth that suggest she was around three years old.

But the rest of the skeleton, especially the foot and knee bones, confirms the ability of these little australopithecines to walk upright in a relatively modern human way but she could not run like more human species. The old adage about being able to 'walk before you run' seems to have had an evolutionary truth to it. If we could see these individuals going about their daily lives, I suspect that it would be very disconcerting – child-sized beings walking about with their child-like bodies but very ape-like heads.

Most interestingly, the new find reveals the presence of a gorilla-like shoulder blade and curved finger bones. Gorillas typically knuckle-walk but they can climb trees as well. These features reinforce the suspicion that Lucy and her kin were still adept at climbing trees. This adaptation was beneficial in a wooded environment with increasingly open grasslands, rivers and lakes occupied by grazing mammals such as extinct species of horse, wildebeest, rhinos, etc along with elephant, suids, giraffe, gazelle giant tortoises along with freshwater species such as hippos, crocodiles and fish.

Our understanding of the diversity of our extinct human related ancestors has grown enormously over the decades but as we have seen, most of the species are only known from the very fragmentary remains of a few adult individuals. The processes by which skeletal remains are recruited to the fossil record tend to discriminate against small bones and juvenile skeletons. This child was probably caught unawares by a flash flood, swept away, drowned and buried in the river flood sediments. Her tragic end is at least allowing us to understand more about her kin and how they were related to us.

Further reading

BAHN, P.G., *The Cambridge Illustrated History of Prehistoric Art*, 1997, Cambridge University Press. (A reference book but readable and well illustrated.)

FOLEY, R., *Humans before Humanity*, 1997, Blackwell new ed. (A readable and authoritative introduction at undergraduate level but assumes background knowledge.)

GIBBONS, A. *The First Human*, 2006, Doubleday, USA. (A well informed account of recent discoveries).

KINGDON, J. *Lowly Origin, Where, when, and why our ancestors first stood up.* 2003, Princeton University Press, USA. (A fascinating an well informed argument from an African perspective).

KLEIN, J. and TAKAHATA, N., *Where do we come from? The molecular evidence for human descent*, 2002, Springer, Berlin. (A fairly technical and detailed account but well written for the informed reader)

KLEIN, R. G., *The Human Career: Human Biological and Cultural Origins*, 1999, University of Chicago Press 2nd ed. (A good university level textbook.)

LEWIN, R., *Bones of Contention: Controversies in the Search for Human Origins*, 1997, University of Chicago Press 2nd ed. (Very useful background reading but slightly out of date now.)

LEWIN, R., *Principles of Human Evolution*, 2003, Blackwell, 2nd ed. (A university level textbook but well written and up to date.)

MITHEN, S., *The Prehistory of the Mind*, 1998, Phoenix (A readable and popular account of the origins of art and religion.)

MITHEN, S. *The Singing Neanderthals: the origins of music, language, mind and body*, 2005, Weidenfeld and Nicolson, London. (An entertaining and informed argument that portrays the Neanderthals in a new light).

PALMER, D., *Seven Million Years: the story of human evolution*, 2005, Weidenfeld & Nicolson, (A detailed popular account)

PALMER, D. *Fossil Revolution*, 2003, HarperCollins. (An account of how fossil discoveries changed our world view.)

STRINGER, C. and ANDREWS, P., *The Complete World of Human Evolution*, 2005, Thames & Hudson. (An excellent introductory university level account).

TATTERSALL, I. *Becoming Human: Evolution and Human Uniqueness*, 2000, Oxford University Press. (An excellent popular account.)

Web sites:

http://www.becominghuman.org/
http://www.encarta.msn.com/encnet/refpages/RefArticle.aspx?refid=761566394
http://www.talkorigins.org/
http://www.leakeyfoundation.org/
http://www.stoneageinsitute.org/

Glossary

words preceded by an asterisk have their own entry

Absolute dating: techniques, such as *radiometric dating, that provide information about age that are derived from physical measurement of the material (see also *relative dating).

Acheulian: an ancient tradition of tool making characterised by *handaxes usually worked on both sides and known as *bifaces. First found at St Acheul in Picardy, France in the early decades of the 19th century but since discovered throughout Eurasia and Africa. The industry appears some 1.5 million years ago and lasted until about 150,000 years ago.

Adaptation: a characteristic of an organism which fits it for a particular environment; also the process by which an organism is modified towards greater fitness for its environment.

Alluvium: old name for surface deposits lying above the *Diluvium subsequently joined together to form the *Quaternary (see below).

Anatomically modern humans: those humans whose anatomy is very similar to that of living humans and who first appeared around 130,000 years ago. There are significant differences between them and *archaic modern humans who lived between about 300,000 and 100,000 years ago.

Anthropoids: a group of higher *primates which include the Old and New World monkeys, apes and humans and originated some 40 million years ago.

Archaic modern humans: early modern humans who lived between about 300,000 and 100,000 years ago and are neither *Homo erectus* nor *Homo sapiens* but evolved perhaps from an evolutionary link between the two such as *Homo heidelbergensis*.

Artefact: any object made, modified or used by humans and ancient human relatives.

Aurignacian: the first major tool industry of early Upper *Palaeolithic age in Europe, from ?40/37,000-31,000 (34-28,000 radiocarbon) years ago, characterised by steep sided scrapers, long retouched blades, bone points and occasionally jewellry and art. First found at Aurignac in southern France by Edouard Lartet with *anatomically modern human remains and bones of extinct animals.

Aurochs/aurochsen: extinct wild cattle (*Bos primigenius*) which were often illustrated by *Cro-Magnon artists on cave walls and carvings and only died out in 1627. Ancestor of today's cattle and were first domesticated about 8000 (6000 radiocarbon) years ago.

Australopithecines: meaning southern ape, a group of early *hominids (of the genus *Australopithecus*) including at least 7 species that lived in Africa between about 4 and 1 million years ago, some of which had very *robust jaws and teeth (also characterized as *paranthropines) and others were more *gracile in form.

Bifaces: see *handaxe.

Biological classification see *classification, biological

Biological species, see *species, biological

Biped: an animal that can walk upright on its two hind legs.

Bovids: a grouping of cattle, sheep, antelope etc which are hoofed mammals with even-toes.

BP: standard timescale abbreviation meaning Before Present (see also *MA*).

Browridge: a ridge of bone above the eyes.

Cha/circumflex/telperronian: a tool industry found in central and southwest France and northern Spain around the boundary between the end of the Middle and beginning of the Upper *Palaeolithic between about 38-34,000 (36-32,000 radiocarbon) years ago. It shows some features of both *Mousterian and Upper Palaeolithic traditions, including some of the earliest bone, antler and ivory objects and is associated with the late *Neanderthals. Named after Cha/circumflex/telperron in central France.

Classification, biological: the arrangement of organisms in related groups based on similarities and differences.

Computerized tomography, (CT): a method of taking X-rays that images successive 'slices' through a body. The scanner is linked to a computer that analyses the output and can produce 3D images of internal structures.

Cro-Magnons: the first *anatomically modern humans in Europe, named after a rock shelter near the town of Les Eyzies in the Dordogne.

Culture: an archaeological term based on very large or widespread collections of material objects such as stone and bone tools that show some common characteristics and belong to the same space-time continuum, such as *Magdalenian. Now broadened to include behavioural traits, belief systems etc and is a more embracing term than industry (see below).

Diluvium: deposits thought to have been laid down by the Biblical flood, found above Tertiary deposits and below the *Alluvium (see also *Quaternary).

Dimorphic (sexually): a condition found in many animals in which one sex, usually the males, is much larger and stronger than the other. Also found in modern humans but in some populations weight differences between the sexes has all but disappeared.

DNA: deoxyribonucleic acid, the self replicating genetic material present in every cell.

Endocast: a cast of the internal space of the skull, closely approximating the shape of the brain.

Flint: a natural, crypto-crystalline silica mineral which occurs as irregularly shaped nodules in certain limestone strata, especially the Cretaceous age Chalk (100-65 million years old) in Europe. Hard and brittle like glass, it can be fashioned into a variety of shapes with very sharp edges and was a favoured material for stone tools.

Genus: a taxonomic category containing one or more species which are each one another's closest relatives. One or more genera are grouped into another category – the family.

Glacial: a cold period during an ice age when ice sheets and glaciers grow and sealevel falls (see also *interglacial).

Gracile: a 'shorthand' term used for a group of australopithecines that had relatively lightly built skulls compared with another group, the *'robust' australopithecines (also known as the *paranthropines) that had much more massive skulls, jaws and teeth.

Grooming: the cleaning of hair by hand or teeth to remove parasites, often practised between individuals to reinforce social bonds.

Hand axe: a stone tool of varied shape from oval to triangular, with both faces worked (also called a *biface) to give a sharp edge, characteristic of the *Acheulean tradition of the Lower *Palaeolithic but also found in some *Mousterian industries.

Holocene: the second and present epoch of the *Quaternary period which began 12,000 (10,000 radiocarbon) years ago and is climatically the *interglacial in which we are now living.

Hominids: members of the extended human family which includes all the species of *Australopithecus, Homo* and the great apes.

Hominines: members of the human family, including the extinct fossil members, the australopithecines.

Homo: the genus to which all humans living and extinct belong, of which between 5 and 8 species of humans have so far been found. They lived over the last 2.5 million or so years and ranged from *Homo rudolfensis* to *Homo sapiens*.

Ice Age: a colloquial term for the most recent of 9 or more prolonged cold phases in the history of the Earth most of which have lasted 20-100 million years. The recent *Pleistocene Ice Age has, so far, consisted of some 24 stages of global climate change oscillating between relatively long cold phases (*glacials) and slightly warmer and shorter phases

(*interglacials) which last tens of thousands of years. Altogether, the recent ice Age has lasted some 2 million years.

Interglacial: a relatively warm climate phase during an ice age when ice sheets and glaciers retreat and subsequently sealevel rises. Global climates are at present in an interglacial phase (see also *glacial).

Industry, stone tool: tools with certain forms repeated and found within a limited geographical region and timespan, which may be a single site or specific region, suggesting that they were made by a single group of people. More narrowly defined than the broader term *culture.

Isotopes: forms of a chemical element that differ in the number of particles in the nuclei of their atoms. Thus they have similar chemical properties but slightly different physical ones. Most elements have two or more naturally occurring isotopes some of which are usually radioactive. Differing isotopic ratios of certain elements and can be used in radiometric dating, proxy measures of climate change and analysis of the environments in which organisms live.

Levallois: an advanced stone tool technology which appears in the Lower *Palaeolithic by 400,000 years ago or earlier and becomes increasingly used from 350,000 years ago, i.e. the beginings of the Middle Palaeolithic of which it is a characteristic. It requires the preparation of a stone core from which thin flakes were then struck, and allows a certain degree of predetermination of the ultimate form of the flakes detached. Named after the Levallois site, now in the Paris suburbs and associated with the *Neanderthals.

Lower Palaeolithic: earliest part of the *Old Stone Age beginning around 2.4 million years ago with the *Oldowan stone tool tradition in Africa and includes the *Acheulean, ending with the development of Middle Palaeolithic flake tools from 350,000 years ago. Some experts link the start of the Middle Palaeolithic with the beginning of the *Levallois technique.

MA: standard timescale abbreviation meaning 'million years' (see also *BP*).

Magdalenian: the main late Upper *Palaeolithic tool tradition in western Europe, dating from about 21-14,000 (18-12,000 radiocarbon) years ago, which includes stone, bone and antler tools, decorative items and many painted caves eg Lascaux. Associated with *anatomically modern humans, it is named after the rock shelter of La Madeleine in the Dordogne.

Magnetostratigraphy: the study of the prehistoric succession of polarity reversals in Earth's magnetic field.

Middle Palaeolithic: characterised by the flake tools of the *Mousterian tradition between around 350,000 and 35,000 (33,000 radiocarbon) years ago; separates the Lower and Upper Palaeolithic.

Molecular clock: a means of timing the evolutionary separation of groups of organisms based on assumptions about

Mousterian: a stone tool industry characterised by certain technologies such as the *Levallois technology as well as the production of small hand axes, triangular points and sidescrapers. Mainly associated with the *Neanderthals and lasting from over 350,000 to around 28,000 (26,000 radiocarbon) years ago. Named after Le Moustier in the Dordogne.

Neanderthals: the most recently extinct human species (*Homo neanderthalensis*) who lived in Europe and Eurasia between about 200,000 and 28,000 (26,000 radiocarbon) years ago and share a common ancestor with modern humans about 400,000 years ago. Recent DNA analysis reinforces their separateness from modern humans and suggests that they did not interbreed with the *Cro-Magnons, with whom they overlapped for some thousands of years.

Neolithic: (New Stone Age) prehistorical phase, succeeding the Mesolithic in Europe, associated with the beginnings of cereal cultivation, cattle and sheep domestication, ceramic pottery and increasing sedentism and village life in the Old World, starting around 12,000 (10,000 radiocarbon) years ago in western Asia.

Oldowan: the oldest know tradition of tool making which began in Africa at least 2.4 million years ago and consists mainly of choppers, flakes and hammer stones. They are generally made from cobbles from which flakes have been struck. Named after Olduvai Gorge in Tanzania where they were first recognised but they have since been found as far away as China and Georgia.

Old Stone Age: see *Palaeolithic.

Palaeolithic: (*Old Stone Age) the first long period of the Stone Age which began around 2.4 million years ago in Africa with the beginnings of the *Oldowan tool tradition and in Europe was followed by the Mesolithic around 12,000 (10,000 radiocarbon) years ago. (see Lower, Middle and Upper Palaeolithic).

Palaeomagnetism: the remnant magnetism of certain minerals within certain rocks. The polarity and orientation of the minerals, relative to Earth's magnetic field when they were originally formed, can be sometimes be measured. Such measures can be usefull for dating the rocks relative to the known magnetostratigraphy and in tectonic reconstructions.

Paranthropines: a group of *robust australopithecines placed in their own genus, *Paranthropus, by some experts. They were originally thought to represent highly dimorphic males as opposed to the *graciles who were thought to be the more lightly built females.

Permafrost: ground in polar and alpine regions which is permanently frozen to depths of up several hundred metres throughout the year apart from the surface which may unfreeze during the short summer. At present some 20$^{\%}$ of the Earth's surface is permafrost but it is diminishing with global warming.

Pleistocene: the first epoch of the *Quaternary period, dating from about 1.8 million years ago until 12,000 years ago, which includes the most recent Ice Age and is characterised by a series of *glacials and *interglacials. The epoch is divided into Lower, Middle and Upper with the boundary between the Middle and Upper dated

at around 128,000 years ago (see also *Holocene).

Point: a category of stone tools consisting of pointed tools flaked on one or both sides, used as knives and, from Middle *Palaeolithic times to tip spears.

Primates: a group of mammals with features in common that includes all humans, apes, monkeys, lemurs, lorises and tarsiers and their fossil representatives who originated at least 65 million years ago.

Quaternary: the youngest system of geological deposits comprising the old divisions of *Diluvium and *Alluvium, now known as the *Pleistocene and *Holocene respectively and laid down over the last 1.8 million years.

Radiometric dating: absolute dating based on the known decay rate of radioisotopes of certain chemical elements including carbon (see *radiocarbon dating).

Radiocarbon dating: a method of dating organic materials such as wood and bone, based on the rate of decay of radioactive carbon isotopes but only effective over the last 55,000 years. Radiocarbon years are not exactly calendar years and so significant adjustments have to be made in converting one to the other.

Relative dating: techniques that provide information about age of fossils, artifacts and sites by indirect means such as correlation with other fossils, artifacts, sites or other information (see also *absolute dating).

Robust see *paranthropine.

Selection, sexual: the mechanism whereby mate choice within a population effects characteristics inherited by their offspring.

Solutrean: a late Upper *Palaeolithic stone industry found in Spain and France between 23-21,000 (21-18,000 radiocarbon) years ago, named after the open air site of Solutre/acute/ in eastern France. The tools are characterised by thin and flat *laurel-leafed* points, often with considerable surface retouching.

Species, biological: a division of a *genus; a group of interbreeding natural populations which do not interbreed with other such groups.

Stratigraphy: the study of sedimentary rock strata, including their processes of deposition, lithification, subdivision and correlation for *relative dating.

Suid(s): members of the pig family, although now reduced to 9 wild species. Distributed throughout Africa and southern Eurasia were previously much more diverse and widely distributed.

Tertiary: a grouping of relatively young geological strata first recognised in the 18th century and now largely replaced by the Paleogene and Neogene. Preceded by the Cretaceous, the last period of the Mesozoic era and succeeded by the *Quaternary. The Tertiary is subdivided into a number of epochs from the Palaeocene to the Pliocene and is now known to have lasted from 65-1.8 million years ago.

Tool industry: see *industry.

Upper Palaeolithic: the last epoch of the Palaeolithic in Europe, North Africa and parts of Asia, lasting from about ?40/35-12,000 (33,000-10,000 radiocarbon) years ago and characterised by diversified blade industries, working of bone, antler and ivory, decoration, art and sophisticated hunting gathering and fishing.

Index

Picture credits

All photographs copyright Science Photo Library except for the following: front cover (l), p13 (t) Skullsunlimited.com; back cover (l), p18, p22, p141 NHPA; back cover (r) p4-5, p6, p88, p90, p90-91, p97, p98, p99, p100, p101, p103 Stephen Morley; p10 Reproduced by permission of the British Geological Survey. © NERC. All rights reserved. IPR/76-34 C; p11, p12 (t), p13 (b), p33 (t), p53 (t, b), p64, p69, p70, p71, p75, p79, p81 Douglas Palmer; p40 Natural; p42 Newspix; p54, p56 South Tyrol Museum of Archaeology www.iceman.it; p71 (b) Teylers Museum, Haarlem, The Netherlands; p73 Bridgeman Images; p85 Thomas Stephan © Ulmer Museum, Germany; p87 University of Bergen; p105 (b) supplied by the University of Southern California; p113 (t) NASA/GSFC/METI/ERSDAC/JAROS,and U.S./Japan ASTER Science Team; p115 Illustrated London News; p48, p117 (t) English Heritage/University College London; p117(b) John Sibbick; p113 Boneclones; p156 (b) Professor Michel Brunet; p160 Camera Press Ltd; p162, p164 UNESCO; p171 City of Westminster Archives.

Every effort has been made to contact the copyright holders.

Maps on front cover, back cover, p20-21, p44, p50-51, p66-67, p92-93, p106-107, p124, p136, p154-155, p166-167 © Martin Sanders. Maps on p38 (t, b), p39, p42, p43, p55, p59, p83, p87, p105, p111, p114, p116, p129, p132, p140, p145, p147 © Bill Smuts.

Illustrations and diagrams on p23, p25, p26, p27, p30, p35, p58, p61, p86, p95, p101, p120, p126, p130, p131, p146, p156, p158, p170, p172, p173, p176, p177, p180, p181, p183, p185 © Debbie Maizels.

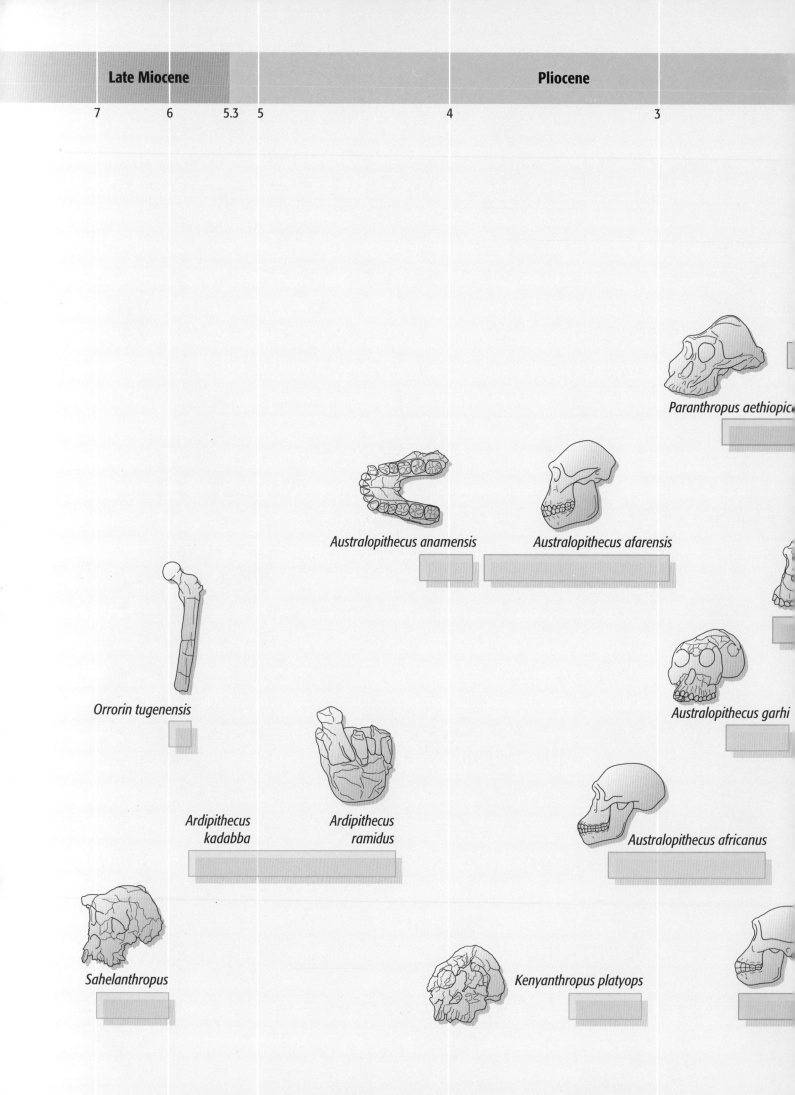

Late Miocene

Pliocene

7   6   5.3 5   4   3

Australopithecus anamensis

Australopithecus afarensis

Paranthropus aethiopic

Orrorin tugenensis

Ardipithecus
kadabba

Ardipithecus
ramidus

Australopithecus garhi

Australopithecus africanus

Sahelanthropus

Kenyanthropus platyops